DAVID L. RUSSELL, P.E.
Engineering Consultant

REMEDIATION MANUAL FOR PETROLEUM-CONTAMINATED SITES

LANCASTER · BASEL

Remediation Manual for Petroleum-Contaminated Sites
a **TECHNOMIC**® publication

Published in the Western Hemisphere by
Technomic Publishing Company, Inc.
851 New Holland Avenue
Box 3535
Lancaster, Pennsylvania 17604 U.S.A.

Distributed in the Rest of the World by
Technomic Publishing AG

Printed in the United States of America
10 9 8 7 6 5 4 3 2 1

Main entry under title:
Remediation Manual for Petroleum-Contaminated Sites

A Technomic Publishing Company book
Bibliography: p.

Library of Congress Card No. 91-67570
ISBN No. 0-87762-876-9

To
K. D. L. who made the writing of this book possible,
J. R. L. who provided invaluable assistance and comment,
and to Marianne, Jennifer, and Laura who made all necessary.

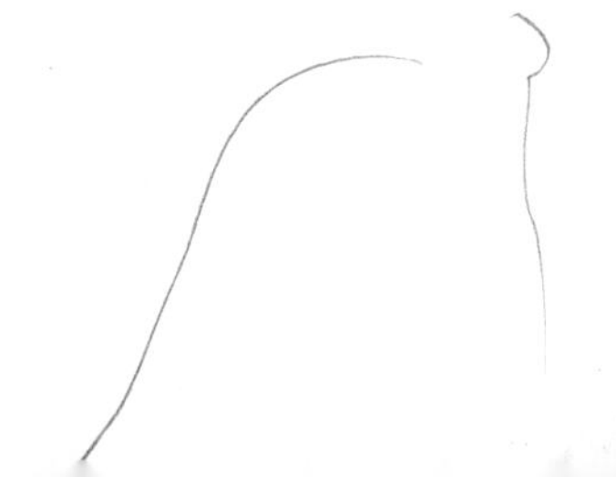

Table of Contents

CHAPTER 1

Introduction and Overview

1.1 INTRODUCTION

THIS manual is designed to assist with the cleanup of gasoline- and diesel-contaminated retail outlets and bulk terminals. This chapter provides an overview to the cleanup process, a brief introduction to the hows and whys of site cleanups, and the management of consultant/contractors which site owners employ. In later sections of this manual, geology, data gathering, cleanup alternatives, and remediation costs are discussed in greater detail. In the following discussion, specific solutions for a typical site cleanup are avoided, as each site is unique, and so are its problems and cleanup solutions.

1.2 REGULATORY FRAMEWORK

1.2.1 CERCLA/SARA

The cleanup of a petroleum-contaminated site is governed by a number of federal and state regulations. The federal law mandating site cleanups is the Comprehensive Environmental Response, Compensation, and Liability Act of 1980. The law, better known as CERCLA or Superfund, provides for the cleanup of all environmental contamination found at any type of location. The law was modified in 1984 by the Superfund Amendments Reauthorization Act (SARA). SARA amended the Resource Conservation Recovery Act (RCRA) and established the Underground Storage Tank (UST) program.

The authority and requirement for cleaning up leaking underground storage tanks comes directly from CERCLA/SARA. The act establishes strict liability for cleanup of an industrial or commercial site for contamination created by any owner(s). The "strict liability" provisions mean that in order to prove that an owner or operator is responsible for cleanup, one has only to prove the owner or operator owned or operated the property and that the contamination has occurred.

The "joint and several liability" provisions of the law mean that any owner or operator with any operations on the site can be included in the list of parties potentially responsible for cleanup. Moreover, the amount of payments made for cleanup will be in accord with the ability of the owner or operator to pay, and not in regard to the amount of contamination that they may have caused.

One of the reasons for private cleanup of gasoline- and diesel-contaminated sites is that it costs less for private industry to perform the cleanups than it does if the government forces the cleanup. A brief example will illustrate this point.

> If the cost of a privately performed cleanup is $500,000, it will be about twice that if performed by a state or by the U.S.EPA. The government body can recover the costs of the cleanup plus a penalty of up to three times the cost of cleanup. The government can also add administrative costs to the base price of the cleanup.
>
> Under this example, the private cleanup costs $500,000. The government cleanup will cost about $1,150,000 including a 15% management and administrative fee. If the government then imposes the maximum 300% fine on the owners of the site, the total charge will be $4,600,000 (threefold penalty plus original cleanup cost). Therefore, it is much cheaper for the private company to perform the cleanup on its own initiative.

1.2.2 SAFE DRINKING WATER ACT

One of the governing pieces of legislation behind the cleanup standards for groundwater and soils is the Safe Drinking Water Act. In its various modifications, it has been law since the early 1970s, and it requires communities to provide safe and clean drinking water to their residents. About 40% of the United States gets its drinking water from the ground, and with the large number of recently publicized groundwater contamination problems, the EPA and the states have reordered their priorities to provide the groundwater with the best protection available. The federal government and many states will not permit the reinjection of any water into the ground from a commercial or industrial source unless that water meets or surpasses drinking water standards.

The federal government has established "health-based" criteria for groundwater. These criteria (known as maximum contaminant levels or MCL's) are related to the number of excess cancer cases per 1,000,000 persons due to the exposure of the most sensitive people (receptors) who are exposed to specific chemicals in groundwater.

The health-based criteria are often applied indiscriminately by the states and by the U.S.EPA, regardless of whether or not there is a receptor population using the aquifer and whether or not the aquifer is potable or brackish (high salt content). Some states have developed their own MCL's for the predominant chemicals found in gasoline. Currently, there are MCL's for benzene only, but MCL's for toluene, ethylbenzene, and xylene are under development.

Other states have merely indicated that they want cleanup to be equivalent to background levels of the contaminants found naturally occurring in uncontaminated soils in the vicinity.

1.2.3 RCRA

The EPA has another weapon to force cleanup of sites. Reporting of spills or "releases to the environment" is mandatory, and there are huge fines and jail terms available to those who choose not to report the releases. However, if a release is not voluntarily cleaned up, the state or the federal EPA can declare that a "disposal" has taken place. When this happens, the provisions of the Resource Conservation Recovery Act are invoked, the site is forced to obtain a RCRA Part B Permit (for hazardous wastes), the cleanup is mandated by a consent decree from the court, and the standards imposed for the cleanup are substantially more stringent and expensive. It is in a site owner's best interests to avoid having to deal with cleanups under RCRA and CERCLA/SARA.

Gasoline or any other compound that has a flash point less than 140°F is considered a flammable hazardous waste. Since the passage of the RCRA Amendments, it has been illegal to dispose of a hazardous waste on land. Gasoline contains a number of compounds for which there are specific hazardous designations in the RCRA.

The new Toxic Characteristic Leaching Procedure test (June 1990) is a means of determining whether or not a waste is hazardous. While there is an exemption for petroleum-contaminated soils and for materials contaminated by petroleum products, a number of state governments are now asking for analyses of contaminated soils by the TCLP test, and may soon seek to regulate these soils as hazardous wastes.

There are only about five landfills in the U.S. which are permitted to accept hazardous wastes. Disposal prices for hazardous wastes are currently around $300 per cubic yard plus transportation, and the total cost for analysis and disposal can easily exceed $1000 per cubic yard.

1.2.4 UNDERGROUND INJECTION CONTROL PROGRAM

The Underground Injection Control (UIC) program governs the disposal of liquids beneath the surface of the ground. While the program was primarily developed to control deep well injection of hazardous waste materials, it has been applied to situations where proposals have been received to permit re-application of waste materials into the groundwater. The principal reason for enforcement of the UIC program is that it forces the applicant to address the concerns a state may have regarding possible aquifer contamination from the injection activity. The UIC permits are expensive to obtain.

Even if water is pumped from the ground and treated, it generally cannot be reinjected, leaked, pumped, or infiltrated back into the ground unless it meets or surpasses drinking water standards, and even then many states are reluctant to allow reinjection because of the possibility of groundwater contamination from the treatment unit. It is possible to add nutrients to the groundwater and still have a potable water well within the drinking water limits.

This does not say that every state will require a UIC permit even to consider the reinjection of treated groundwater. There are states that are actively modifying their UIC program regulations to permit limited reinjection of water, as may be required for biological treatment of contaminated aquifers.

The use of soil-flushing solvents has been widely promoted in the waste and water treatment literature and in a limited number of laboratory and field trials. It appears to be a promising technology. However, many of the flushing agents are potentially harmful to humans, and may be difficult to remove from the soil completely. Under these circumstances, it is doubtful that many of the states will approve a soil flushing chemical any time in the near future.

1.2.5 OTHER WATER REGULATIONS

Other water regulations which will be encountered in the establishment of a groundwater cleanup program include local sewer ordinances and stream discharge standards. Before any commercial establishment can discharge wastewaters or wastes into public sewers, a permit must be obtained from the local municipal authority. Even if the municipality is lax about permitting the discharge and the need for a permit, it is best to provide the authority with written notification of the discharge and then an analysis or characterization of the waste stream in a timely manner *before* the discharge occurs.

The discharge of any treated water or process water or wastes to a storm sewer or directly to a stream must be permitted by the state or the federal government. Failure to get a permit can result in fines of $10,000 per day and personal liability to anyone who authorized the discharge. The discharge permit is not generally hard to get, but sometimes the state may impose monitoring requirements on the discharge to determine whether or not water quality standards and toxicity standards are being met. Generally the discharge of treated waters to a storm drain will fall into the class of a minor permit if the total flow is less than 50,000 gallons per day. If the proposed discharge is over this quantity, the permit acquisition may be much more involved, and qualified consultants or the company's environmental affairs department should be contacted to assist in obtaining the permit.

1.2.6 AIR REGULATIONS

The installation of treatment equipment for cleanup of the water at a petroleum-contaminated site may require several different types of air permits. These permits are often overlooked by field personnel and their consultants.

The first regulations concern the odor control statutes. If the underground venting system or stripping system has a discharge of materials into the air, it will produce an odor. Depending upon the location of the facility and the proximity of neighbors and the amount of material discharged to the air per hour, the installation may cause odor problems. These odor problems can be serious, and their potential should not be ignored. Odor problems can cause the odor-emitting facility to be declared a public nuisance and can lead to closure of the site and fines for the company operating the facility.

The second group of regulations concerns the emission of air toxics. Benzene, while it represents only about 1.5% of the total volume of gasoline, is a highly regulated carcinogen (cancer-causing agent). Gasoline contains a wide variety of organic compounds, a number of which are subject to regulation as potentially "toxic air pollutants." Before anyone constructs a facility that can emit gasoline vapors to the air, the state and county health and environmental control agencies must be contacted to find out whether the emission of gasoline is regulated and how it is regulated.

The third class of air pollution control regulations includes those that regulate the volatile organic compounds that cause air pollution and ozone. Of itself, gasoline is classified as a volatile organic compound (VOC), as are over 70% of the components (by weight) that comprise gasoline. Volatile organic compounds cause air pollution and breathing problems when the compounds oxidize to ozone. Each state has air pollution control regulations that govern the emission VOC's; depending upon how the state has established its air pollution control districts, the size, location, and enforcement powers of the local air control districts may vary widely. In some areas the county is the enforcing air control agency, while in other areas the air control enforcement district may encompass several counties, or the state may retain all permit authority.

Some states absolutely limit the total amount of pollutants discharged, and others have a sliding scale that permits a lower total discharge of pollutants in an urban area and a higher discharge in a rural area. Either way, the discharge point will be treated as a point source and will require a permit. If the proposed quantity of VOC's emitted is above the limit, an air pollution control device may be required. The air pollution control device, generally a flare or a condenser, will be a permitted treatment unit, and it will require a construction permit and an operation permit.

It is necessary to determine which air regulations may be applicable and to secure the necessary permits *before* construction is begun on a facility.

1.2.7 CONSTRUCTION AND OPERATING PERMITS

If pollution control or treatment equipment is to be installed, a permit is generally required. Often the permit is waived or is covered by the remedial action plan, describing the corrective actions that are required to restore the site. The permitting requirements for ancillary equipment vary from state to state. If a discharge is made to a storm sewer, the state will insist on a construction and operation permit. The treatment equipment may need the services of a licensed wastewater treatment plant operator, an operating engineer, or another person technically trained and licensed by the state to operate a treatment facility. Also, the person must have received the required OSHA training for hazardous wastes or hazardous chemical/material handling as required. The minimum OSHA training required for anyone handling chemicals is specified by 29CFR 1910.120.

1.2.8 FIRE, HEALTH, AND SAFETY REGULATIONS

All construction impacting the health or welfare of the public must have a building permit and must meet certain design and life safety codes. These codes are reflected in the structural, electrical, building, and fire codes, and in the material standards for equipment such as those issued by the American Society for Testing Materials (ASTM), the American Concrete Institute, American Steel Institute, the National Electrical Code, and so on. Installation of surface and underground storage tanks, for example, is covered by National Fire Prevention Association (NFPA) Code number 30 and 30A. Failure to follow that code in the design or construction of a service station can, in the event of an accident, subject the owners of the site to liability for damages, and possible criminal liability for negligence.

When a retail outlet is built or modified, a registered professional engineer or architect is retained to insure that the local building and fire codes are followed. Unfortunately, there are no written codes or standards of practice to follow in the design of an underground remediation system.

Consequently, there are a large number of firms processing sites without proper attention to good engineering and safety practices. When good safety and engineering design practices are followed, the opportunities for fire and accidents are minimized.

It would be in the best interest of anyone dealing with a remediation design to insist that NFPA 30A and other standards and local building codes are followed and that a registered professional engineer is retained to inspect and seal the plans for the design of the remedial system.

The remediation systems should also be maintained by persons who are knowledgeable in the handling of hazardous materials and who have had the required OSHA training.

1.3 MULTI-MEDIA PROBLEMS

It is easy to get petroleum into the ground, but hard to get it back out again. The degree of difficulty encountered in cleaning up a petroleum-contaminated site depends upon the site geology, hydrology, mineralogy, the size of the spill, the amount spilled, and the length of time since the spill occurred. The problem does not exist in the groundwater or in the soil alone, but in a dynamic balance between the soil, water, and the air in the soil. Gasoline diesel fuels and any of their volatile components can and do move between the soil, the air, and the water.

A recent study of a 30,000 gallon gasoline spill indicated the relative magnitude of the problem. Sixty-two percent of the volume of the spill was contained in the soil in a free phase. Thirty-two percent of the volume was adsorbed onto the soil. The balance, estimated from 1%-5% of the total, was in the water. However, when the volume of contaminated soil was considered, the free phase contaminated about 7100 cubic yards, or about 1% of the total. The adsorbed gasoline contaminated about 150,000 cubic yards, or about 20% of the total. The dissolved phase contaminated approximately 960,000 cubic yards, or about 79% of the total soil [1].

Soils are produced from weathering of the bedrock or decaying vegetation, or are deposited by water or wind. Soil is composed of many types of silica and aluminum silicate minerals, and it also contains organic carbon. Each of these components forms a dynamic system that has its own distinct chemistry, surface properties, and physical characteristics. A number of soils act as natural adsorption systems for oil and gasoline. Organic chemicals can attach themselves to soils with a fierce tenacity, which makes the removal of the chemical very slow. Soils with silts and clays are more difficult to clean up than those which contain only sands.

Clay and silt soils are difficult to clean up because the mineralogy of the soil gives it surface-attractive properties that make the petroleum cling to the surface of the soils. The attraction of the surface combined with the small particle size causes a little clay to be able to hold a lot of chemical. For example, a one foot cube of clay or silt with an average particle diameter of 0.0005 cm has a surface area of approximately 135,400 square feet, or 3.11 acres. Even if the adsorption rate is a few grams of material per square foot, a little clay goes a long way.

Sand, by comparison, has a larger particle size and it has a mineralogy that is predominantly silicon oxides. As such, it is much less reactive and surface-active than the silts and clays with their aluminum mineralogy.

The effect of particle size on soils and the difficulty of removing contaminants can be seen in the work of Wilson and Brown. In a recent study, the adsorptive capacity of sand, one of the easiest soils to remediate, has been estimated to be 32 grams of gasoline per kilogram of soil (g/kg) for coarse sand, and 122.36 g/kg for fine sand [2]. Wilson and Brown also discovered that simple washing of the soil with water is not effective in removing contamination. Flushing a soil with 46 volumes of water removed only 1.6% of the contamination. Even after the contaminated soils were flushed with 500 pore volumes of water, the contamination levels were still at 1400 mg of gasoline per kilogram of soil.

Cleanup of contaminated rock formations is extremely difficult because the petroleum can adhere to the rock just as it does to the soil, and it can get into fissures where it is difficult or impossible to remove with conventional technology.

Any spilled petroleum product will partition itself between the soil and the water and the air in the soil. The undissolved product (free product) floats on the top of the groundwater table. The groundwater level in the soil is not static, but rises and falls with the seasons and it is directly affected by the overall water budget for the site and the rainfall/drought cycle in the area. As a result, the groundwater table fluctuations tend to smear the free product through the soil, displacing it physically when the water table is high. At times when the water table is low, there may not be any visible free oil or gasoline in the groundwater. When the water table rises during the winter and spring, it will displace some of the gasoline in the soil and a free product will reappear.

There is never only one way to clean up a particular site. The choice of the method of cleanup will be one of the greatest factors affecting the cost. Another important cost factor is the level of cleanup required to satisfy the regulatory community. The selection of the optimal method for cleanup of a particular site should always be based upon the total cost of the cleanup, considering the time required for the cleanup, the ability of the technologies to reach the desired treatment level, the monitoring cost, the maintenance costs, the strategy for regulatory compliance, and the cash flow requirements for the cleanup option under consideration.

1.4 CLEANUP STRATEGIES

In dealing with a contamination problem, one cannot address just the water or just the soil. Both must be cleaned up. The current focus on vapor stripping and its attendant successes may be somewhat short-lived, as vapor stripping does not remove any but the most volatile components of gasoline, leaving the semi-volatile and non-volatile compounds for other forms of treatment.

The end requirements for cleaning up a site are often negotiable. While a number of states will insist upon attainment of drinking water standards in the groundwater, a number of states make it clear that they address site cleanup on a case-by-case basis. Some states will permit the development of asymptotic limits that permit site cleanup to be curtailed when the removal efficiency of a treatment system reaches negligible levels. Unless regulations that specify the cleanup limit exist, the decision on how clean a site may be is based upon what the regulator will accept.

Many of the regulators are inexperienced and unfamiliar with cleanup standards, methods, and the practical limits to the technology. With this lack of practical experience, there is often little support for any position other than the most extreme—cleanup to the lowest level possible, or detection limits.

1.5 CORRECTIVE ACTION PLANS (CAP)

As a contaminated site is investigated and reported, the state or federal EPA will require a corrective action plan to be submitted. The corrective action plan contains a summary of the site conditions, the contamination, and a proposal and commitment by the owner or operator to reduce the contamination by specific methods.

All corrective action plans have a number of elements in common. These elements are (1) plan certification; (2) summary of extent of contamination; (3) description of corrective actions taken or in progress; (4) a statement of the goals or results of the corrective action; (5) summary of the design and operation of the equipment required for the corrective action system; (6) a description of how the work will be accomplished along with a schedule for implementation for the work; and (7) a plan for detecting changes taking place at the site, and for monitoring progress and compliance with the objectives set forth in the plan.

Some corrective action plans are more elaborate than others, depending upon the requirements of the state. Most, however, follow the federal guidelines as outlined in Federal Regulations Chapter 280. Of the states in the Southeast, Florida is the most elaborate, as their guidance document on preparation of corrective action plans is over 100 pages long and costs $50.00.

In addition to the Corrective Action Plan, states may also request a Quality Assurance and Quality Control Plan for data—before the site characterization work is undertaken. The plan requirements can be extremely detailed, setting forth all applicable sampling protocols, equipment, investigation tools, data and chain of custody management protocols and equipment decontamination and waste disposal procedures. Again, Florida is one of the most stringent states in the Southeast, and their QA/QCP can often exceed 200 pages of detailed instructions.

A corrective action plan does not always have to specify action and compliance monitoring forever. Depending upon site location, the amount of contamination, and site factors, some sites can be closed with little or no additional work. Sometimes a state may require only monitoring, or may be content with the removal of the free product from the site.

A typical corrective action plan outline is shown in Table 1.1.

All corrective action plans require the signature of a responsible corporate officer. CAP's should only be signed by persons who can commit the necessary funds and actions to the project, despite what may be authorized in the corporate budget for the year. Often, the signature of a vice president of an operating division may be required.

The states have the legal authority to impose criminal penalties, including prison terms on the persons signing the plans for the corporation. While the states were initially reluctant to impose these fines, there have been several recent court cases where criminal sanctions are being sought against the vice president of a company who failed to implement the action required in the Corrective Action Plan.

The CAP should be prepared by the remediation consultant or contractor at the conclusion of the remediation and prior to beginning the work on the remediation phase of the project. No remedial work should be begun unless

TABLE 1.1. Contents of the Corrective Action Plan.

Section and Title	Description
1. Introduction	Outlines the location of the site.
2. Certification	Confirms that the plan is in compliance with state regulations.
3. Description of the extent of contamination	Complete description of the extent of the contamination at the site including types of contaminants, groundwater standards, maps, and cross sections showing the areal and vertical extent of the contamination at the site, the location of any free product in contact with the water table, and the extent of the plume of dissolved contaminants.

TABLE 1.1. (continued).

Section and Title	Description
	The description should also include the direction and movement of groundwater, the lateral extent of ground and soil contamination, and the seasonal fluctuation of groundwater levels. Some states may also require equipotential lines to be drawn showing the flow-lines for groundwater and contaminant migration. The regional geology and hydrogeology must be described in detail. Boring logs from explorations must be submitted along with an evaluation of calculations of the hydraulic conductivity, storage coefficient, velocity, and direction of the groundwater flow. Some states will also require the investigation and identification of public and private groundwater wells in the area in the vicinity of the contamination site. Georgia, for example, requires location and identification of all private drinking water wells within a 0.5 mile radius of the site, and all public drinking water wells within a 3.0 mile radius of the site.
4. Objectives of the corrective action	A statement of the desired objectives to be reached.
5. Design and operation of the corrective action systems	A narrative report with appropriate documentation and drawings of the type of system to be used to correct the problems encountered. If air permits or water or sewer permits are required, they should be identified here, or there should be a statement that the permits have been or are being prepared.
6. Implementation of the corrective action program	A schedule with compliance dates for appropriate actions and milestones for accomplishments should be submitted here. This section of the plan should describe how the plan is to be implemented, how the equipment is to be serviced, how frequently it is to be inspected and maintained, what sampling protocols are to be used in monitoring groundwater, what test methods are to be used, what criteria are to be used to judge the effectiveness of the cleanup activity, and which actions will be taken when the site is judged to be clean. Finally, there should be a site closure plan.

the state has reviewed and approved the CAP and the company has received a letter from the state approving the actions. Since some states are just now requiring CAP's, and since the work on some sites is ongoing, the letter from the state must also address past actions taken by the company and grant approval for them.

The natural desire of the company to have the state approve its cleanup actions is understandable, especially when the state may be paying the bill for cleanup under a state trust fund, but in a number of instances the needs of the site may be immediate. Many states have a six-month or longer backlog in approving cleanup payments and actions. In any number of instances, the owner/operator of the site must take necessary actions in a timely fashion and hope that the state will approve of and pay for those actions after the fact. It will be in the owner/operator's interests not to wait for state approval where there are long delays. In a six-month period, a contamination problem can become substantially worse, and the cleanup costs may become much greater because of the spread of contamination.

1.6 MANAGEMENT OF THE CONSULTANT/ REMEDIATION CONTRACTOR

A number of oil companies utilize subcontractors who have skills in engineering, environmental sciences, geology, and hydrology. Individual contracts are authorized under a broad set of specifications that cover such items as site investigation and drilling. The contracts are often executed on a time and material basis, but occasionally they are competitively bid to a fixed price.

1.6.1 TIME AND MATERIAL CONTRACTS

The time and material contracts usually contain provisions that are most favorable to the contractor, which is the equivalent of letting the fox guard the hen house. Even if the contractor makes a mistake, the owner or operator employing the contractor pays for it. Under the contract provisions for a time and material contract, there is often a disincentive to save money. The contractor's fees are in direct proportion to the number of man-hours put on the project and the amount of money spent, because the contracts generally include a surcharge for project supplies. If the owner/operator of the site and the contractor fail to establish a man-hour estimate and total project budget, the owner/operator may be in for an expensive surprise.

In order to use a remediation contractor effectively, the owner or operator must know what is desired and expected before the contract is initiated, and the contractor should be required to produce a proposal that outlines what is to be furnished and at what total price. Additionally, the contractor must not be allowed to have a blank check on key decisions. Some contractors have been known to exaggerate the requirements for site cleanup and to describe conditions accordingly in the corrective action plan documents.

A time and material contract is an acceptable form of contractor management if the owner or operator of the site has the time to manage the contract adequately and to review the contractor's work on a periodic basis and provide instruction when and where necessary. An alternative way to manage a remediation effort is on a fixed price basis with a clearly written set of specifications and bid documents on a project. Competition is one of the most effective forms of cost control.

1.6.2 FIXED PRICE CONTRACTS

One of the most efficient ways to manage a remediation effort is through a fixed price contract. The contractor should be told what specific performance is required and what standards must be met. The financial portion of the contract should include allowances for changes in the contract on some agreeable basis.

The contract should contain specific provisions for safety requirements, equipment standards, materials of construction, codes and standards, workmanship, and applicable laws and regulations that must be followed. Depending upon the specifications and the work to be performed, the contract can be quite simple or may run many pages. Even where the contract is fixed price, the project still needs to be managed and inspected. The role of the inspector is to insure that the agreed upon work has been provided and that the material and equipment quality specified has been installed. The payment terms under this type of contract are generally based upon a percentage of completion, with the owner retaining 10% of the payment to insure completion of the contract. Performance bonds that insure the completion of the contract are usually required on this type of project and are a good idea.

1.6.3 CONTRACTOR MANAGEMENT

Consultants and remediation contractors don't necessarily like having their work and methods questioned. However, it is just this type of contractor micromanagement that may be required to insure that the contractor is doing only what is necessary and not wasting money. Frequent communication with the contractor is often necessary if cost overruns are to be avoided. Examples of contractor excesses are easy to find. Is it really necessary to install a stainless steel screen when you are looking at organic chemicals? Chances are that PVC will do quite well, and the minor degradation that will be encountered can be accommodated at a much lower cost.

If a contractor indicates that there is only one "best" way to perform a specific remediation, or if the field supervisor or manager has questions about the costs or technical feasibility of the proposed solution, the manager would be wise to get a second opinion from another contractor or consultant. The threat of independent review of a contractor's actions will often cause the remediation contractor to improve his internal project and contract management, and this threat will also cause the contractor to be more careful in the performance of the work.

1.7 SPECIFICS OF INITIAL EXPLORATION

All inspection and contract supervisory personnel need to know exactly what the remediation contractor is proposing and that he is accountable for his actions and activities at each step of the investigation. In the initial exploration phase for a site investigation, the contractor should be required to prepare a detailed work plan describing what he will do and how he will do it. The work plan should be site-specific and should include the details as outlined in Chapter 2 of this manual. The preliminary site investigation plan should describe the site completely before any drilling or other invasive work is performed. The preliminary site investigation plan should be performed within the fixed price of the contract and should be reviewed and approved by the field supervisors and the regional engineers before it is accepted. The contractor should not be allowed to begin work on the site until after the initial exploration work plan has been approved.

The work plan should also contain a health and safety plan for the site. The health and safety plan will insure that only qualified persons who have received the necessary training will be permitted on the site, around the drill rig during explorations, or anywhere near operating treatment equipment, and it will insure that the contractor provides adequate protection for his employees.

The work plan should also contain a sampling plan and an analysis plan for the site. The sampling plan should set forth the manner in which the samples of soil and water will be collected, preserved, and analyzed. The contractor should collect enough information to permit several remedial alternatives to be investigated. If the contractor is to perform a vapor survey on the site, the inspector or the field supervisor should determine that the vapor survey equipment conforms to the contract specifications. (Some vapor survey equipment uses a hydrogen flame which may not be permitted under company safety procedures.)

1.8 SPECIFICS OF PHASE TWO INVESTIGATION

The detailed second phase investigation of the site begins when the first draft of the preliminary investigation report has been received for review. The second phase investigation requires closer contractor management than the first phase investigation because it is more specific and the contract specifications will, of necessity, be less detailed. The supervisory personnel in charge of contractor activities need to meet frequently to determine what work was performed under the contract and what the specific findings were. Among the items that need to be discussed between the supervisor and the contractor are the different remedial options that might be feasible and the data requirements that will enable the evaluation of the suitability of a particular solution.

As early as possible in the second phase investigation, a draft corrective action plan should be prepared and reviewed by the supervisor. The purpose of the CAP should be to determine data sufficiency, the need for additional data, the possible remedial alternatives, and the costs of obtaining the additional information. The draft corrective action plan should incorporate several different remedial alternatives that are capable of meeting the state requirements.

The contractor should be instructed to prepare a detailed work plan and sampling for the phase two investigation. Re-submission of the phase one sampling and analysis plan is not sufficient. Even if the contractor has a strong feeling that the best alternatives for the site may be vapor stripping and other appropriate technology, at least one additional alternative should be considered to prevent premature forcing of a technological solution.

The data-gathering activities that the contractor undertakes should be sufficient to develop a full evaluation of the options under consideration at the conclusion of the phase two study and to fill out the corrective action plan completely.

The information in Table 1.2 is a brief checklist of the materials that should be developed and presented at the conclusion of a phase two investigation. At the conclusion of the phase two investigation, the contractor should be able to answer the questions regarding the system alternatives and costs. He or she should be able to set forth a proposed remedial action plan and be prepared to defend it.

The contractor should be instructed to determine the costs and scope of the phase two activities within plus or minus 15% for a fixed price inves-

TABLE 1.2. **Checklist for Phase Two Investigations.**

Item	Description
1.	Complete description of physical characteristics of the site
1a.	Maps, diagrams, charts describing the site and its surroundings
1b.	Topographic mapping of the site showing location of principal structures and location and elevation of wells and utility lines (overhead and underground) and surrounding structures
1c.	Site zoning and infrastructure showing local highways and access (from highway department or city roads and streets department drawings)
1d.	Location of adjacent public and private wells
1e.	Location of other potential sources of contamination in the vicinity which may have a direct impact on the site or which may be contributing to the site contamination
2.	Description of site geology
2a.	Complete description of regional and local geology
2b.	Maps and geologic fence diagrams showing locations of borings, and the depth and type of formations present at the site (preferably using the Unified Classification System)
2c.	Maps and diagrams that show the location and variation of the groundwater at the site
2d.	A description of the groundwater quality and summary of the chemical data for groundwater
2e.	Maps and fence diagrams that show the extent of the contamination in the soil and in the groundwater
2f.	Maps that show the areal extent of the contamination above and below the groundwater table

TABLE 1.2. (continued).

Item	Description
3.	Description of the site hydrology
3a.	Principal description of the regional hydrology and water budget for the site
3b.	Determination of the hydraulic conductivity, storage coefficient, and permeability for each of the principal aquifers as determined from slug tests or pumping tests
3c.	Description of the direction of movement and rate of travel in each of the principal formations on the site
3d.	Rate of movement of the contaminants in each of the principal aquifers
3e.	Maps showing the equipotential flow lines for each of the principal contaminant zones and aquifers
3f.	Results of soil venting tests, including equipotential levels of air vacuum
4.	Chemical characteristics and treatability investigations
4a.	Summary of chemical and physical characteristics of the soils
4b.	Description of soil and water treatment alternatives
4c.	Projected time required for cleanup under different alternative methods of remediation
4d.	Projected total costs of equipment, maintenance, sampling, and operation for each treatment system and cleanup alternative to a common time base
4e.	Annualized total costs for each system considering equipment replacement
4f.	Permits required to operate and maintain the treatment system, and types of operators required for proper operation
4g.	Design and engineering costs for the system and technology constraints, if any
4h.	Estimated construction and installation costs for the systems under consideration
4i.	Description and rationale behind the selection of the preferred method of treatment, including calculations and preliminary design sketches
4j.	Recommended alternative for the treatment of the contamination problem
4k.	Discussion of any public health impacts or problems from operation of the treatment system

tigation. The contractor should, by the time the CAP is finalized, be in a position to determine total project costs for the principal remedial solution proposed and for all alternative solutions. The contractor should include projected manpower needs, the cost of weekly or monthly visits, the cost of sampling and compliance monitoring, and the cost of equipment replacement and maintenance. Because each of the different remedial solutions proposed will have different costs and a different time base for the completion of the remediation, the contractor should provide a comparison of total costs for each system considering the maintenance and monitoring costs, equipment replacement, and interest costs to a common time base. The U.S.EPA recommends that present worth figures be used to determine the overall costs of their remediation efforts, using an interest rate of 10%. For example, if incineration is selected as one remedial alternative, it may have a substantially higher initial cost than pumping and treating solutions. However, when the maintenance and monitoring costs for years of pumping and treating are considered, the cost of incineration may compare favorably.

The contractor should prepare at least two remedial alternatives that contain complete cost information. Each of the alternatives should be able to accomplish the objectives of the state regulations. Depending upon the state rules and the flexibility of the state in setting remedial levels, a final decision on which alternative to adopt may be delayed until the different remedial alternatives have been discussed with appropriate state personnel.

A rehearsal should be held beforehand to discuss what will be proposed and how it will be presented to the state. The negotiation strategy is to present alternatives that will address those items of state concern, but which will be as cost effective as possible. The strategy for presentation should be flexible but should be structured to present the least cost remedial alternatives first and incremental options beyond that point.

Some states, particularly Louisiana, require the presentation of at least two remedial alternatives. Many states are so busy with the work already in hand that they will not approve a remedial alternative in less than six months.

The field supervisor needs to be aware that a remediation activity that is paid for by the owner or operator may have different requirements than one that is being paid for by a state under a reimbursement plan. When the state pays the bill, it may establish competitive bidding rules and will effectively take over the remedial activities at the site.

1.9 REMEDIATION DESIGN AND CONSTRUCTION

At the conclusion of the second phase investigation, the contractor has prepared the corrective action plan and is ready to begin the specifics of the design of the remedial action system. Some states are requiring a remedial action design to be prepared by a professional engineer; they also require the submission of plans and specifications before the remediation actions are allowed to begin. Table 1.3 lists some specific items that may be applicable to a remediation design.

Before the construction can begin, all state and local permits must be obtained. Some of these permits may require months of review time by the state. In many instances, however, temporary permits to construct can be issued while the permit review process is underway.

Some of the permits that may be necessary include a sewer discharge permit, or a state permit to discharge to a storm sewer or navigable waterway; an air pollution control device construction and operation permit; and an Underground Injection Control permit. Some states, such as South Carolina, may require permits to construct withdrawal wells and or monitoring wells.

Local construction and building permits must also be obtained. These permits are issued by the state or local fire marshal's office after the plans are reviewed, and the fire marshal is satisfied that the system will meet National Fire Prevention Association codes, and local building and construction codes.

A professional engineer or architect should be retained to design the remedial system and submit plans and specifications to the state, whether it is required by the state or not. The architect or engineer will be familiar with the requirements for safe design to prevent fire and explosion and will follow the building and life safety codes.

At the conclusion of the project, the contractor's engineer or geologist will be asked to certify that there has been a closure to the site that is in conformance to the state rules. It is often better for a company to have the contractor or the consultant make that certification rather than to have the company supervisors or the regional managers do so, as the contractor is making the evaluation regarding compliance with the law.

1.10 CLEANUP ALTERNATIVES AND SYSTEMS

1.10.1 INTRODUCTION

Cleanup of a site cannot be performed only in the soil or only in the groundwater. This section will provide a brief summary of the most popular alternatives currently being used to clean up the groundwater and the soil at gasoline-contaminated sites. More detail and a discussion of the limitations of each system are provided in subsequent chapters.

In considering the cleanup of a site, one should remember that no single technology works every time and that more than one technology may be required to clean up the site. For example, the technology of vapor stripping

TABLE 1.3. Specific Design Considerations.

Items and Description	Comments
1. Remediation wells—placement	The placement of the collection wells or trenches should be such that they intercept the plume boundary at its greatest extent or such that by manipulation of the hydrogeologic regimen the plume is pulled back toward the well field so that recovery is inevitable.
2. Well-pump compatibility	The materials used to construct the well pumps should be compatible with gasoline. Brass, bronze, aluminum, certain plastics, and stainless steel are suitable. Materials that can cause sparks (cast iron and steel) should be avoided unless explosion levels are checked and the pumps will be continually under water.
3. Centrifugal well pumps	Centrifugal well pumps will make an emulsion of the water and the gasoline. The emulsion may be more difficult for some treatment systems to remove effectively. For high head (lift)-low volume locations these pumps are not generally suitable, as they require circulation of cooling water in the well to keep the pump from overheating. Pipe friction is a factor in the selection of a centrifugal pump system. If non-submersible pumps are used, suction lifts (monitoring well depths) greater than 20 feet may be a problem. If the total dynamic head of the pump is exceeded, the pump will not move water at all.
4. Positive displacement well pumps	If the pump is air-driven, someone should be checking to see where the air is vented and whether or not explosive mixtures of air and gasoline vapors will be created inside the wells. Air-powered positive displacement pumps have low mechanical efficiency. Electrically powered pumps are more efficient. Positive displacement pumps are best used in formations with low permeabilities which are too deep for recovery trenches.
5. Pump and well manholes	Manholes are often considered confined space entry areas under OSHA. Where manholes contain equipment that must be serviced, a minimum diameter of 36″ is often recommended.

TABLE 1.3. (continued).

Item and Description	Comments
6. Electrical controls in the wells	Gasoline vapors in monitor well manholes could accumulate until they exceed the lower explosive limit. Electrical fittings and controls should be explosion-proof.
7. Pipelines from wells to treatment systems	The lines should be laid out for easy detection of leaks and line replacement if leaks occur.
8. Vacuum extraction systems	Check the upper and lower explosion limits (U.E.L. and L.E.L.) for the vapors in the venting system lines during vapor extraction to insure that the system is above the U.E.L. or below the L.E.L. on start-up* Make sure that all motors and controls adequately address the potential for explosion by accumulating gasoline vapors.
9. Air stripping towers (for removal of gasoline from water)	These towers are subject to plugging and freezing. High iron and dissolved mineral levels in the water will create maintenance problems. The towers need to be properly anchored or guyed to prevent their being blown over in a wind storm.
10. Carbon adsorption systems—air	Carbon adsorbers for air require a special type of carbon that is more expensive than many other types. Carbon adsorption systems have a high maintenance and replacement cost. They should not be used unless the volume of air is very small or loading rates are very low. Water vapor in the system may compete with hydrocarbons for adsorption sites on the carbon. See discussion in Chapter 3.
11. Carbon adsorption systems—water	Carbon adsorbers should only be used where flows are very small (less than 5 gallons per minute). Carbon is expensive to replace. See discussion in Chapter 3.
12. Incinerators, catalytic combustors, and flares	Gasoline and air are an explosive mixture. If the feed concentrations into the combustion chamber approach the L.E.L., a flame arrestor and a water safety seal should be used. Catalytic combustors have a specific temperature and fuel loading rate at which they operate. If they exceed these rates or temperatures, the catalyst will be damaged. They should have safety interlocks and alarms to shut them down in the event of malfunction.

(continued)

TABLE 1.3. (continued).

Item and Description	Comments
	Incinerators can operate just as economically as catalytic combustors if the fume incinerator uses a heat recovery device.
13. Product collection tanks—general	By any other name these are called gasoline storage tanks. They need to be treated as such. Proper ventilation, (tank vent heights and configurations) fencing, "Flammable—No Smoking" signs, and property line setbacks need to be observed.
14. Planning—for aboveground tanks	If by any chance the total aggregate storage volume above ground exceeds 1320 gallons total or 660 gallons in a single tank, the facility will be required to have a spill prevention countermeasure and control plan as required by 40CFR 112. It is best to avoid this if possible by keeping aboveground storage requirements below 1320 gallons.
15. Monitoring and collection wells—general	Stainless steel is generally unnecessary for wells where gasoline is a problem. PVC will suffer some minor but acceptable swelling and degradation. PVC is substantially less expensive than stainless steel. For wells in gasoline there are a number of options including stainless wire wrapping on the well screens or use of galvanized wire wrapping on the screens. As long as metals are not the issue of the groundwater investigation, galvanized screens are acceptable. The selection of the "appropriate" well screen materials and screen slot sizes may affect the entire recovery program.
16. Monitoring and collection wells—size	The cost of drilling a 4″ diameter well is not substantially greater than that of drilling a 2″ diameter well. Smaller diameter wells are more difficult to use in groundwater pumping and recovery operations.
17. Controls and systems—general	Treatment equipment and maintenance systems require periodic maintenance. The use of some automatic or remote monitoring systems can cut down on the frequency of maintenance visits at a greater capital cost. Ask questions and get price comparisons for savings.

TABLE 1.3. (continued).

Item and Description	Comments
18. Autobailers	Autobailers are used where formation permeability is low. Autobailers convey the liquid to the top of the well casing, where it must then be pumped to a treatment system. Autobailers are often trailer-mounted and may present a traffic obstacle.
19. Autoskimmers	Autoskimmers remove free product. They are trailer-mounted devices. They work on a buoyancy principal, which raises and lowers a bailer on a cable. Anything that affects the buoyancy of the bailer (bio-fouling) will decrease the efficiency of the skimmer. The sensitivity of the equipment is within plus or minus 1/4″ of free product.
20. Air sparger systems	Air sparger systems are an alternative method of providing aeration to strip volatiles from groundwater. The aeration takes place in a tank rather than in a tower. Bio-fouling is often eliminated.

*As a practical matter a system should *never* be operated above the L.E.L.

and of *in-situ* bioremediation are currently in vogue and hold much promise for economical removal of gasoline contaminants to very low levels. However, if heavy metals from leaded gas are also an item of concern, neither technology will remove metals.

Before a decision is made to select a particular cleanup technology, the total costs and the impact of the cleanup method on the site and upon the community should be identified. Many of the factors that will go into making a specific site decision may be internal to the owner and operator of the site and are not in the purvey of the remediation contractor. The remediation contractor should not be permitted to develop the project schedule and make financial commitments to the state without prior approval from the owner or operator of the site.

1.10.2 DEFINITIONS FOR SOIL AND GROUNDWATER REMEDIATION SYSTEMS

This brief section is designed to describe the basic terms used in developing a remediation system for soil and for groundwater. Many of the terms are used in a manner that applies to the treatment of both soil and groundwater. The treatment technologies defined in Section 3.2.2 are specific to groundwater.

1.10.2.1 Definitions for Soil Remediation Systems

The following technology descriptions may be useful in developing an basic understanding of the appropriate methods used to clean up petroleum-contaminated sites and groundwater. These technology descriptions are broadly applicable. For specific information on each of the technologies, the reader is advised to consult Chapter 3 of this manual.

1.10.2.1.1 *Biological Treatment*

Biological treatment is the aerobic ("with oxygen") or anaerobic ("without oxygen") process of using microorganisms to consume the hydrocarbons in the soil or in the groundwater. The process can utilize naturally occurring soil or animal microorganisms or naturally or artificially adapted microorganisms from other sources.

In practice there are three types of aerobic treatment in general use and two types of anaerobic treatment in less frequent use. The treatment can be classified by the type of container or the location of its application. Contaminated soil that has been excavated or that is quite shallow is generally treated by application of water, fertilizer, and manure to the soil. The soil is treated in lifts of less than a foot in thickness, and is thoroughly tilled and turned. This type of treatment is often referred to as *land farming* or *land treatment*.

Biological treatment is most successfully used in the treatment of contaminated groundwater. The process is similar to that used by many municipalities in the treatment of sewage by the activated sludge process, in which the contaminated water (with nutrients added, if required) is subjected to sustained aeration for up to 48 hours. The liquid is then subjected to quiescent settling to remove any bacteria. The bacteria are returned to the process for further feeding. In anaerobic treatment, a period of sustained mixing is substituted for aeration. Anaerobic treatment is effective in removing some less volatile compounds but is frequently found to be temperature-sensitive and difficult to operate.

In-situ *bioremediation* takes place in the undisturbed, unexcavated soil or groundwater. The process can be aerobic or anaerobic. The carbon source for the bacterial growth is the organic carbon and petroleum products in the contaminant plume. Nutrients are supplied to the soil by pumping them in the groundwater, or by trickling them in from the surface. In aerobic *in-situ* bioremediation the oxygen is supplied by direct aeration of the soil, by hydrogen peroxide, or by other chemicals.

1.10.2.1.2 *Containment/Sealing*

Containment and sealing of the soil is the placement of a barrier to the movements of the contaminants. For soil contamination, the containment can include surface sealing by asphalt, concrete, or the construction of a specially designed surface cap that includes membrane liners. Lateral migration is often prevented by construction of a slurry wall or other barrier around the edge of the contamination source. Truly effective containment designs will include consideration of a natural barrier (unbroken rock or thick natural clay deposits) at the bottom of the containment walls. Surface sealing will reduce the downward potential for movement of the contaminants by reducing the downward movement of water through the soil.

1.10.2.1.3 *Encapsulation/Solidification*

The encapsulation and solidification processes are closely related. Neither is effective in the treatment of contaminated groundwater. Both rely on mixing of the soil with cement, asphalt, silicate, or plastic to form a solid product that resists the release of the contaminant by modest abrasion, pressure, crushing, or chemical attack. The acceptability of a solidified or encapsulated material can be determined by laboratory procedures that subject it to moderately acidic conditions and determine the quantity of contaminant released after a given test period.

Specific chemical formulations are effective only against a limited number of chemical compounds. Cement and silicate formulations are generally not effective in solidification of gasoline-contaminated soils. Asphalt contaminant systems can be effective in containing petroleum products, but that may also be due to the fact that they mix the gasoline and fuel oils with asphalt materials that have the same source and greater concentrations. The process is generally used in connection with excavation and is seldom performed *in-situ* on contaminated soils.

1.10.2.1.4 *Vitrification*

Vitrification is a relatively new process; not much is known about its long-term viability. The process converts soil into a silicate matrix by electrical melting in place. The heating of the soil is accomplished by burial of carbon electrodes and resistance heating of the soil through generation of electric currents or through microwaves. Because the melting temperatures of clay are in excess of 1500°F, the process may not be suitable for active service station sites. Economic viability of the process is yet to be confirmed.

1.10.2.1.5 *Groundwater Controls*

Groundwater controls are generally thought of as the manipulation of the direction and rate of movement in the groundwater by raising or lowering the groundwater table at a specific site. Groundwater flows in the direction of decreasing hydraulic gradient (downhill), and the process is often compared to re-contouring the groundwater table and surface profiles on the site. The purpose of the manipulation of the hydraulic regimen is containment or collection of the groundwater by control of its movement across the site.

1.10.2.1.6 *Incineration*

Incineration is the high-temperature thermal oxidation and processing of a soil. The process generally destroys all organic matter in the soil. It is not effective in the destruction of inorganic materials such as lead, chromium, zinc, or other metals. State air permits are required to operate an incinerator. Before the soil can be processed, it must be excavated and screened. The process is not recommended for active service station sites because of space requirements and the presence of a source of combustion (the incinerator burner). Some states, principally Kentucky, Tennessee, and South Carolina in the Southeast, recommend the treatment of petroleum-contaminated soils by incineration or by incorporation of the soils into an asphalt batch plant feed stream.

1.10.2.1.7 *Pump-and-Treat Systems*

Pump-and-treat systems are combination systems in which the water and fuel or gasoline mixture is collected by using wells or collection trenches, and is pumped to a treatment system. If a free product layer is present, two pumps may be used per well; one of those pumps will be used to remove the free product and the other will be used to depress the water table, thereby increasing the thickness of the free product layer and making recovery easier.

Alternatively, one pump may be used to pump both petroleum and groundwater. A single pump may be used where the concentration of free product petroleum is low, but it can be used to pump both gasoline and water even if the thickness of the free product layer could warrant a two pump system. If a single pump is used, the gasoline is removed and separated for recovery by a treatment system.

Treatment of the water to remove gasoline may be accomplished by biological treatment, aeration (stripping), or by carbon adsorption. Specifics of water treatment systems must be tailored to the chemistry of the water at each site. If air stripping is used, thermal processing may be required to reduce hydrocarbon concentrations to the air.

1.10.2.1.8 *Reinjection*

Reinjection is the re-use of treated groundwater at a site for the purposes of promoting the management of groundwater movement, the enhancement of biodegradation, or the elevation of the water table. Because of the sensitivity of the states to the potential for groundwater contamination, reinjection is not generally considered unless it can be shown to have a definite advantage. Where reinjection is proposed, a state may require an Underground Injection Permit to be obtained and may also require extensive groundwater monitoring. Treated groundwater is often required to meet drinking water levels before it can be reinjected into the ground. The treatment to these levels is often uneconomic.

1.10.2.1.9 *Soil Washing*

Soil washing is the physical removal of contaminants by flushing the soil with water, detergents, solvents, or nutrients. The process may occur as a by-product of bioremediation or may be performed to enhance the bioremediation process. The emphasis is on the physical removal of the contamination from the soil by the action of the washing liquid. Soil washing may be conducted *in-situ* or on the surface. It is difficult to achieve in silty and clayey soils, and the process has limited effectiveness when used as a remedial option by itself.

1.10.2.1.10 *Vapor Stripping*

Vapor stripping is the removal of volatile materials from the soil and groundwater by the process of inducing or forcing an air movement through the soil. Vapor stripping will effectively remove the most volatile fraction of the gasolines in a soil but will not remove it all. Vapor stripping is less successful on diesel and fuel oil contaminated soils because they contain relatively small proportions of volatiles. Vapor stripping may also have some benefit in the treatment of a contaminated groundwater plume, but it is not an effective technology when used solely for contaminated groundwater. Vapor stripping is generally accomplished by installation of a vacuum extraction well in the soil zone immediately above the groundwater table (vadose zone).

1.10.2.2 Definitions for Groundwater Treatment Systems

Technology descriptions for groundwater treatment are partially tied to the method used to pump the groundwater to the treatment system. Some of the terms defined below will inevitably be encountered in discussing groundwater remediation systems. Chapter 3 discusses some of the techniques, pumping systems, and limitations of various groundwater treatment technologies.

1.10.2.2.1 *Air Stripping*

Air stripping is the process of utilizing the vapor pressure difference between gasoline and air to extract gasoline from water with air. Given enough time and contact with air, many of the dissolved components in gasoline will evaporate into the air. Air-stripping systems come in many shapes and sizes. The most common shape is that of a narrow cylinder between 10 and 30 feet tall. The stripping towers are filled with plastic packing that has a high surface-to-volume ratio. The water is introduced through the top of the tower and air is blown upward through the falling water. The countercurrent stripping system is the most effective, but other systems can achieve equal treatment and removal levels with minimal increases in total project costs.

Depending upon the amount of air and water treated, the stripper may have a mist eliminator to reduce the droplet carryover. Droplets from a stripping tower may be a nuisance because they can cause spotting of cars in the area.

The diameter of the tower and the tower height are determined by the levels of volatile compounds in the water, the quantity of water to be treated, and the treatment level to be attained in the effluent. Tower heights are generally less than 25 feet; tower diameters are sized depending upon air flow requirements but are usually less than 2 feet in diameter.

Iron and high mineral content in the groundwater can form deposits that will plug the tower. If waters are high in iron, a different type of stripper system may be used to achieve the same results.

1.10.2.2.2 *Biological Treatment*

The description provided above in Section 3.2.1 is applicable to this section as well.

1.10.2.2.3 *Blower Systems*

A centrifugal blower that uses straight-sided vanes around the rim of the enclosed center flywheel to displace the air is often referred to as a regenerative blower. These blower systems are characterized by relatively low pressures and vacuums (generally under 12″ of mercury or pressures up to about 7-10 psig) with quiet operation. Many of the blower housings and rotors are made from aluminum. The all-aluminum configuration may be considered explosion-proof, but the manufacturers do not warrant it as such.

While rotary blowers may be adequate for some systems, the requirement for high air capacity and high vacuum may dictate that the blower be a positive displacement lobed or vaned blower. Both types of blowers can generate significant pressures and vacuums. Many of these blowers are quite noisy and will require silencers and motor enclosures.

1.10.2.2.4 *Carbon Adsorbers*

A carbon adsorption system consists of a tank or vessel that has water distribution piping in the top, a bed of activated carbon or charcoal, and water collection piping in the bottom. The tank may or may not be a pressure vessel. Smaller carbon systems use 55-gallon drums with PVC piping systems. Almost all carbon adsorption systems operate at low pressures, generally under 20 pounds per square inch.

Activated carbon is not very efficient in removing dissolved gasoline from water, and it requires a lot of carbon to treat the water effectively. The typical removals of gasoline are measured in the milligrams of gasoline removed per kilogram of carbon used (parts per million by weight). Carbon systems usually require the installation of a pre-filter to remove sediment. See the discussion in Chapter 3.

1.10.2.2.5 *Chemical Precipitation*

Chemical precipitation is a system of removing dissolved minerals from water, which usually requires the addition of lime, alum, or ferric chloride. These systems are occasionally used in the treatment of petroleum-contaminated waters, but they produce a sludge by-product that must be removed and disposed of in a hazardous waste landfill. This is not recommended for groundwater treatment systems. Chemical precipitation is not effective in the removal of dissolved organic materials from water.

1.10.2.2.6 *Coalescers*

The root meaning of coalesce is "to bring together." That is exactly what a coalescer does—it brings the suspended or emulsified drops of oil or gasoline together in water to allow them to hit each other and grow in size until they are big enough to float to the surface of the coalescer or separator tank, where they are skimmed off. Coalescers are usually used in connection with an API (American Petroleum Institute) separator. The system provides a baffled area with closely spaced submerged plates that are set at a 45° vertical angle to the flow, or a larger area that contains something resembling an oversized Brillo pad. The coalescer area is followed by a quiescent settling area, where the coalesced droplets can come together at the surface for removal by skimming.

Another type of coalescer is a cartridge-style system that resembles a large industrial thread cone. While this type is different in design, the principle of operation is the same as that previously described. The coalescer is discussed in Chapter 3.

1.10.2.2.7 *Enhanced Bacteria*

The use of bioengineered or enhanced bacteria for the treatment of gasoline-contaminated water or soil is not encouraged. The bacteria may initially show some promise in the laboratory and in some field conditions, but the genetic mutations that are bred into these "super bugs" are frequently lost within several days or weeks of their initial application.

It is far better to provide nutrients and oxygen to the soil and let Mother Nature take her course. Bacteria in the soil will naturally develop enzymes that will enable them to degrade the compounds in gasoline, once the concentration of gasoline in the water falls below the toxicity level of the bacteria.

A more thorough discussion of treatment alternatives, their advantages, and their limitations is provided in Chapter 3.

1.10.2.2.8 *Ion Exchange System*

An ion exchange system is a bed of specially prepared plastic beads that have engineered properties that remove anions or cations from water. Depending upon the levels of dissolved minerals and gasoline in the water, ion exchange can be used to remove the iron, which creates problems with a stripping tower. The ion exchange system requires a periodic regeneration of the ion exchange bed. Commercial services are available in many parts of the country. While not used for the direct treatment of gasoline-containing waters, it may have some use in related water treatment areas.

1.10.2.2.9 *Permeability*

Permeability is a measure of the ability of an aquifer to permit the movement of water through it under a unit hydraulic gradient. Gravel soils have permeabilities of 10^{-2} centimeters per second. Very impermeable clays have values that are less than 10^{-6} centimeters per second. Alternatively, permeability is measured in units of gallons per day per square foot.

1.10.2.2.10 *Transmissivity*

Transmissivity of an aquifer is the value of permeability multiplied by the value of the thickness of the aquifer. The units must be consistent, and the measurement relates to the level of specific yield from an aquifer.

1.10.2.2.11 *Vapor Condensing Systems*

This is a manner of controlling vapor emissions from a vapor venting system by chilling the plume until the gasoline vapors condense out of it. The refrigeration equipment chills the exhaust temperature down to about −70°F. These systems are not widely used because of the high capital cost and small return of free product for the investment.

1.10.2.2.12 *Vapor Incinerators and Catalytic Combustors*

Vapor incinerators and catalytic combustors are high-temperature thermal oxidation devices. The catalytic combustors have a limited mass loading rate and a lower organic destruction efficiency than incinerators. Incinerators, unless they are equipped with a heat recovery system, are not as thermally efficient as a catalytic combustor because they operate at a higher temperature. Vapor incinerators have a destruction efficiency that can exceed 99%. Catalytic combustors have a volatile organic compound destruction efficiency that ranges from 95% to about 98%. The installation of either will require an air pollution control permit.

1.11 SOIL TREATMENT

The treatment of soil must not be separated from the treatment of groundwater. Table 1.4 presents several of the more widely used soil cleanup and treatment systems. The information will enable the reader to evaluate the effectiveness of the treatment technology for the cleanup of soils by themselves. The systems listed in the table carry a dual ranking, one for effectiveness of the treatment system and one for the ease of implementation of the system. Additionally, the table is broken into two parts to permit evaluation of solutions that are more applicable at inactive sites (closed or remodeled stations) and those that are more effective at

TABLE 1.4. Suitability of Remediation Alternatives: Contaminated Soil Treatment Site Use.

Type of Soil	Surface Bio.	*In-situ*	Contain.	Encapsul.	Excavate	Classific.	Incinerate	Process	Wash	Vapor Strip
	Remediation Alternative (Inactive)									
Gravel	f3	f3	p3	g1-f2	g1	p3	g3	g1	g1	g1
Sand	g1-f2	g1-g3	p2	g1-f2	g1	p3	g2	g2	f1	g1
Fine sand	g1-f2	g1-f3	p2	g1-f2	g1	p3	g2	g2	f2	g1
Silty sand	g2-f2	g2-f3	f3	g2-f3	g2	f3	g2	f3	f3	g2
Silt and clay	g3	g2-f3	f2	f3-p3	g2	f3	g2	f3	f3-p3	f2
Clay	g3	g2-f3	g2	p3	g2	f2-f3	g2-g3	f3	p2	f2
Very tight clay	g3	g3-f3	g3	p3	g2	f2-f3	g3	f3	p3	f3
Perched water table	g3-f3	g3-f3	f2-g3	p3	g3-f3	f2-f3	g3-f3	f3-p3	p3	f3
Lens structure in soil	g3-f3	g3-f3	f3-p3	p3	g2-g3	p3	g3	f3	f2-f3	f2-f3
	Remediation Alternative (Active)									
Gravel	Same as	f3	Same as	u	Same as	Unsuited	Unsuited	Unsuited	g1	g1
Sand	above if	g1-f3	above but	u	above only	high	open	unless	f1	g1
Fine sand	volumes	g1-f3	may be	u	if volumes	temp.	flame	volumes	f2	g1
Silty sand	small and	g2-f3	difficult to	u	are small	process	and space	are very	f3	g2
Silt and clay	soil moved	g2-f3	perform	u		u	required	small—	f3-p3	f2
Clay	offsite	g3-f3	because of	u		u	u	requires	p2	f2
Very tight clay		g3-f3	excavation	u		u	u	space	p3	f3
Perched water table		g3-f3	required	u		u	u	u	p3	f3
Lens structure in soil		g3-f3		u		u	u	u	f2-f3	f2-f3

Success of treatment: g = good; f = fair; p = poor; u = unsuited.
Ease of accomplishment: 1 = good; 2 = fair; 3 = difficult.

active sites. The evaluations are made with respect to the types of soil at the sites.

The overall ranking in the tables is as described below.

- success of treatment
  - Good: Treatment will successfully decontaminate the soil, leaving an acceptable level of residual contamination or none at all.
  - Fair: Treatment will generally decontaminate the soil, but will leave higher levels of residues or may leave pockets of contamination that will require more work and expense to achieve a good level of treatment.
  - Poor: Consider something else. The treatment technology is marginally suited or unsuited to the task at hand. As an example, soil washing is a poor choice for clayey soils because the clay is not a permeable material.
- ease of accomplishment
  - 1: Easy to accomplish. This technology is readily understood by many practitioners in the field, and contractors are generally familiar with the techniques at hand and can efficiently and effectively perform the work at minimum cost. Examples include excavation of a soil at an inactive station. All the contractor has to do is dig up the material and haul it away.
  - 2: More difficult to accomplish (and more expensive to implement and perform). Contractors are slightly familiar with the solutions, and they have greater difficulty in accomplishing the task. Examples include the excavation of silty and clayey soils as compared to sandy and gravelly soils.
  - 3: Quite difficult to accomplish. The contractor can accomplish the solution with greater difficulty and more expense. Examples include excavation of a soil where a perched water table exists (groundwater may require pumping or de-watering in order to excavate), and *in-situ* bioremediation in very tight clays where the technology is difficult to accomplish because it is difficult to get water or air through those clays to provide nutrients to the organisms therein.
  - u: Unsuited, use a different technology. Either it will not work or it is totally unsuited to the application at hand because of the physical factors associated with the construction of the option under consideration.

1.12 GROUNDWATER TREATMENT

Often the criteria by which the success or failure of the treatment system, and ultimately the remedial option, are judged are the treatment levels attained by the water treatment system, and the length of time that is required to clean up the site. The failure of the treatment equipment to attain removal levels, or the failure of the remedial collection system to remove the contaminants from the ground in a timely fashion, can cause the entire remediation system to be judged as a failure. The failure of the treatment system and the failure of the collection system are not related to each other by anything more than the fact that the two systems are connected by piping.

The water treatment system is designed to remove petroleum products from the water and produce an effluent that can be re-used or that can be sent to a sewer for further treatment by a publicly owned treatment works (POTW). If the treatment equipment is meeting its design conditions and if it is producing an effluent that contains the desired treatment levels, it is probably performing properly. If the volume of water being treated increases or decreases substantially, the treatment system may need to be modified.

The extraction of the contaminants from the ground and cleanup of the site is a problem in geology, hydrology, and engineering. There are a limited number of things that can be done to clean up groundwater at a gasoline contamination site, and all of them are severely restricted by the geology of the site.

The first problem to be considered in treating groundwater is a problem in geology and hydrology. What is the extent of the plume? What are the concentrations of the BTEX and other contaminants in the plume? In which types of soil do the contaminants reside? The answers to these questions will place constraints on the development of a groundwater treatment system.

The presence of a free product phase may also influence the selection of the groundwater treatment system. A decision to recover the product in the ground will force the designer to use either a two pump system or some type of coalescer system that will separate the gasoline from the water in the treatment train.

If the discharge level required of the treatment system is in the low parts per billion range, suitable for return to the groundwater or for direct discharge to a stream, the designer's options are limited to aeration and carbon adsorption. If, however, the treatment system effluent requirements are in the parts per million range, then a number of other options are available. These options include biological treatment without additional treatment: spray towers, chemical oxidation, carbon adsorption, and partial aeration of the wastewater stream.

Vacuum extraction systems help remove volatile organics from the soil and help evaporate the free product. They do not, however, directly help remove contaminants from the groundwater except in that by removing the largest part of the soil contamination, they eliminate the problem of recontamination of the groundwater by the soil.

TABLE 1.5. Suitability of Remediation Alternatives: Groundwater Contamination Treatment Site Use: Active or Inactive.

	Remediation Alternative										
Type of Soil	Surface Bio-remediation	*In-situ* Bio-remediation	Contain-ment	Encap-sulation	Excava-tion	Classifi-cation	Incinera-tion	Direct Aeration	Stripping Tower	Vacuum Stripping	Carbon Adsorption
Gravel	g1/g1	g1/g1	Not	Not	u	u	u	g3/g1	g1/g1	Not for	g1/g1
Sand	g1/g1	g1/g1	a	a	u	u	u	g3/g1	g1/g1	water	g1/g1
Fine sand	g1/g1	g1/g1	treat-	treat-	u	u	u	g3/g1	g1/g1	treatment	g1/g1
Silty sand	g1/g1	g1/g1	ment	ment	u	u	u	g3/g1	g1/g1	removes	g1/g1
Silt and clay	f1/f1	f1/f1	option	option	u	u	u	f3/f1	f1/f1	free	f1/f1
Clay	f1/f1	f1/f1			u	u	u	fp3/fp2	fp1/fp1	product	fp1/fp1
Very tight clay	fp1/fp1	fp1/fp1			u	u	u	p3/fp2	p1/p1	from soil	p1/p1
Perched water table	fg1/fg1	fp1/fp1			u	u	u	f3/f1	fp1/fp1		fp1/fp1
Lens structure in soil	gp1/gp1	gp1/gp1			u	u	u	f3/f1	fp1/fp1		fp1/fp1

Success of treatment: g = good; f = fair; p = poor; u = unsuited.
Ease of accomplishment: 1 = good; 2 = fair; 3 = difficult.

Soils become much more difficult to treat as the amount of silt and clay in the soil increases and the permeability of the soil decreases. Many hydrocarbons in gasoline and diesel fuels, if present in high concentrations, will tend to have a dehydrating influence on the silt or clay and cause it to shrink and crack, much the same as if they had been placed in the summer sun to dry. When this occurs, it creates a natural pathway for the movement of the gasoline. Replacing the gasoline with water may cause the clay to swell, restoring its integrity but making the movement of the petroleum and water through it much slower.

Table 1.5 lists some of the more prevalent treatment options and ranks them with regard to their effluent quality and effectiveness, and it considers the permeability of the soil in which the problem exists. The information in the table is presented in a manner similar to that of Table 1.4, but the variables are slightly different. In Table 1.5, there are two sets of numbers: the first set addresses the ability of the treatment system to achieve an effluent in the parts per billion range; the second set of numbers addresses the ability of the treatment system to achieve an effluent concentration in the parts per million range—one suitable for discharge to a POTW.

It should be noted in the previous table that the difficulty of implementing a particular system of water treatment increases with the amount of silt and clay in the soil.

1.13 SUMMARY OF TREATMENT OPTIONS

While there is no one treatment system that is right for all locations, some of the technologies, such as vapor stripping and bioremediation, are suitable for a wide range of conditions, and their application will make sense in a majority of the locations. Vapor stripping and bioremediation are not, however, universal panaceas for all service station remediation sites. There is no one solution that will work for all types of problems. Each solution must be tailored to the specific requirements of the site and the regulations of the state.

The cleanup of contamination at service station sites will require months and years of work, and many tens of thousands of dollars of expense at each site. Because the remedial technology is so dependent upon site-specific geology and other conditions, it is impossible to predict accurately the success of a technology for a particular site.

If the contamination occurs predominantly in the ground, the groundwater is shallow, or the station is in need of reconstruction, it is better to consider direct methods of removing the contamination and incinerating or replacing the soil with clean material. In the long run, this is one of the cheapest technologies, and it is the most effective, as it removes the contaminant immediately.

The most effective treatment for groundwater appears to be biological treatment. If the soils are predominantly sands and gravels, it may be wise to consider a combination of surface biological treatment and *in-situ* biological treatment. If the soils are predominantly clays and silts and the permeability is low, *in-situ* biological treatment should be strongly considered.

The most effective manner of collecting the groundwater is in trenches. The next most effective method for collection is by pumping wells.

The most effective method of removing free product from the soil and the surface of the groundwater appears to be a combination of pumping wells and vacuum extraction. While these methods are not universal, they appear to have a wider range of applicability than many other technologies or combinations thereof.

In a sandy soil, pump-and-treat methods may prove very effective in removing most of the contamination. However, to remove the last of the contamination from the soil and from the groundwater, it would appear that the best technology available is *in-situ* biological treatment.

For groundwater in soils that are predominantly clays and silts, the most promising treatment appears to be biological treatment. The difficulty of manipulating the groundwater in these soils cannot be underestimated. The success or failure of the biological treatment will depend upon the ability of the remediation contractor to get the nutrients into the soils at the right place and for a long enough time.

REFERENCES

1 Wilson, S. B. and R. A. Brown. 1989. "*In-situ* Bioreclamation: A Cost-Effective Technology to Remediate Subsurface Organic Contamination," *Ground Water Monitoring Review*, 9(1):614-671.

2 Wilson and Brown, op. cit.

CHAPTER 2

Data Requirements

2.1 INTRODUCTION

THE most important task in the development of a site assessment or in the preparation of a remediation plan is the collection and analysis of the data. This data must include accurate information about the geology and hydrology, as well as the type, concentration, and areal extent of the contamination.

Insufficient or poor information can create failure of the remediation effort before the onset of the activity, wasting manpower and capital. Some of the ways in which data collection effort can fail include: (1) poor plume definition, which causes incomplete cleanup; (2) inexact definition of the geology, causing selection of the wrong remediation method; (3) inexact or improper collection of chemical data, causing selection of inefficient or inappropriate cleanup methods to be used. These failures will add additional cost to the cleanup. At the extreme, mistakes in the interpretation of the data can cause failure of the remediation effort.

At different times in the life of a remediation project, different information will be required. When the initial investigation is begun, the data-gathering activity has a limited scope, focused mainly on the determination of whether or not contamination exists. Later, after the contamination has been verified, the focus of the investigation shifts to determination of the nature and extent of the contamination. Later still, before the remediation can begin, the investigation is focused on the specific application of a few technologies, and it seeks to determine the applicability of those technologies to the contamination problems at the site.

Often, the original site investigation contractor or a subsequent contractor must repeat some of the initial work with a specialized purpose. However, if the data gathering by the initial investigator is sufficiently detailed, the second effort will be less repetitive, more effective, and less costly.

In the development of site and characterization information, the investigator must balance the needs of the project against the cost of development of the data. The most complete method of developing data on the site would be to dig up the entire site, although that is impractical and expensive. Balanced use of monitor wells, vapor probes, test pits, and surface geophysics may offer the most effective method of getting information about the site.

The final outcome of the remediation effort cannot be determined while the exploration is underway. However, with a clear idea of the types of remedial solutions and applicable strategies, the investigator can tailor the development of data toward those ends. The more complete the investigation is, the better and easier the choice of cleanup methods will be.

This chapter will discuss the general type and kinds of information needed for a more complete and thorough site assessment, and the data requirements for several of the most common types of remediation efforts – pump-and-treat, bioremediation, vapor stripping, and incineration.

2.2 GENERAL SITE INFORMATION

The following list of general information about a site should be presented in the initial site characterization report, and that report should be submitted to the client before any subsurface geology is performed. A listing of initial site data that should be contained in a preliminary assessment is shown in Table 2.1.

The purpose of the preliminary assessment information gathering effort is to present as complete a picture as possible of the site and its surroundings and the subsurface environment in which the contamination may exist. Many of the information sources listed in Table 2.1 are self-explanatory. A few words of information about other sources may be helpful.

Street and utility drawings should be obtained to provide a more complete picture of the subsurface environment. Retail locations are often found in areas where the installation of storm or sanitary sewers may have had a particular influence on the local hydrogeology. Most sewers are bedded in and backfilled with sand and gravel. The sewer bedding may serve to provide a conduit for the movement of contamination as it did in the spread of contamination from the Love Canal. Generally the sewer bedding has a permeability of between 10^{-1} to 10^{-3} centimeters per second. The water movement through a typical gravel or stone sewer bedding could be on the order of 1000 up to 100,000 times greater than many clays found throughout the East and Southeast.

Sewers are laid on a slope with the natural terrain, but occasionally against it. Contaminated groundwater has been shown to travel along sewer bedding for substantial distances, sometimes over a mile or more. The

TABLE 2.1. Preliminary Assessment Site Data.

Information	Source
Location maps	
U.S. Geological Survey topographical map, 7.5 minute quadrangle	U.S. Geological Survey, and most state geological survey offices and commercial map supply stores
City or county street map	State, city, and county highway departments
Property tax map	Tax assessor's office
Survey plat of the site showing property boundaries	Tax assessor's office
Street and utility drawings, including buried electrical and telephone lines, as well as drawings of storm and sanitary sewers and construction standards	District or township engineer's office, or city engineer's office, and gas and telephone utility offices
Aerial photographs of the site and its surroundings	Defense Department or Department of Agriculture
Soil Conservation Survey site map for soil conservation	County Soil Conservation Service Office
State Geologic Survey map of subsurface conditions	State Geologic Survey
"As-built" drawings of the site showing the location of the underground tanks, and the location of buried lines and utilities	Available from site owner for most newly built locations and for installations where the tanks have been replaced
A listing and summary of the geologic information on public and private water wells within a three-mile radius of the site	Available from the State Geologic Survey, or from the groundwater protection unit of the local state
Location of other underground storage tanks within a three-mile radius of the site	Available from the underground storage tank office of each state. Information is limited to only those facilities that have registered their tanks.
Any state-ordered cleanup, CERCLIS Site information, or any available information on remedial activities conducted on properties within a three-mile radius	Available by Freedom of Information Act request from the U.S.EPA and from each state office

sewer bedding materials may also serve as a barrier to development of effective groundwater recharge of a particular area in which remediation efforts are being attempted.

The Soil Conservation Service has a substantial body of information about soils throughout the United States. While the SCS Soil Survey Maps are available almost everywhere, their coverage is not universal. For example, Rockland County Georgia does not have a completed SCS map. The survey has been completed, but the map is still under preparation. For a specific site, however, the soil information is available by contacting the County SCS office.

Each soil in a soil map is classified by type. The SCS Map has extensive information on the soil types in the upper 5 feet. The information includes depth to seasonal high water table; composition of parent and daughter soil classifications, and depth of each type of profile; approximate soil classification by USDA, Unified, and AASHO classification systems; approximate soil sieve size and screening information; liquid and plastic limit data; approximate vertical permeability; and structural and foundation information. Additionally, SCS also provides a discussion of the general characteristics and suitability of the soils for certain types of activities, principally related to farming and home-building. The SCS survey information can provide much useful information about the types of soils in the area, and can give the geologist a useful starting place for beginning the assessment work.

A typical soil series map and description for a location in Cobb County, Georgia is provided in Appendix C for informational purposes only.

Aerial photography is available for most of the United States through the National Cartographic Information Center and through the U.S. Geologic Survey, both of which are in Denver, Colorado. The CIC telephone number is 303-236-5829; the USGS telephone number is 303-236-7477. The photography is available in negative form or in print form. The National Archives has also been a consistently good source of historic information, providing old aerial photographs of sites across the U.S. Index sheets are available for each year and area photographed from both agencies and both are modestly priced. Photographic information may require up to about 6 weeks to obtain.

At one time aerial photography was available from the Department of Agriculture, and each county has recent aerial mapping available from the county Soil Conservation Service Office under that program. Historical photographs of a specific area are available through the National Archives.

Wide-area photogrammetric coverage of a region is also available from satellite photography. While this information can be obtained from the EROS Data Center, Sioux Falls, South Dakota, 57198, it may not be of much use for individual site investigations because the scale and resolution of the satellite photos may not provide enough detail for practical use in remediations at a single site. The largest practical resolution is one meter or more in diameter.

The State Geologic Survey for each state has detailed maps of the subsurface geology and rock formations under principal sections of its state. Many states have also compiled a list of principal aquifers and wells within the state. Georgia, for example, has a survey of all private and public wells in the state between 1940 and 1980. Some states such as South Carolina have passed public registration laws for water well installation, requiring well drillers to provide the state with a boring log and well completion diagrams.

One of the many reasons for inclusion of well information around a particular site is that several states, including Georgia, South Carolina, North Carolina, and Virginia, are evaluating the cleanup objectives at a site in terms of the proximity of neighbors and the potential for human exposure to the chemicals at the site. In Georgia, North and South Carolina, and Virginia, the presence of public wells within a state-defined radius of up to three miles may cause the state to seek more stringent cleanup requirements.

A listing of underground storage tanks within a one-mile radius of the site should identify the presence of otherwise hidden potential sources that may be contributing contamination to that aready at the investigation site.

Information about environmental contamination on adjacent properties can sometimes be gained by examining the property transfer assessments of recent sales if these records are available. Unfortunately, these investigations are not available except through the lender, or unless the site has had contamination that was reported to the state. When available, the property transfer assessment reports generally are not uniform in format or in quality, but they can serve as a good indicator of the likelihood of local site contamination.

Physical examination of open trenches and excavations in the area around the site can often be an inexpensive source of information about site subsurface conditions. These examinations should be undertaken while the geologist is in the area preparing for the startup of the initial boring program.

The preparation of a preliminary site assessment will help guide the geologist and the owner in developing a feel for the potential of the site, and help them make initial inferences about the subsurface conditions to be encountered. The geology of a specific site must be complete and make sense in the context of the regional geology.

2.3 SITE EXPLORATION PROGRAM

If the subsurface exploration is to be successful, it must be flexible, addressing the many possibilities, and the exploration must be adequately supported. The geologist needs to have adequate supplies and personnel at his disposal in order to insure a successful outcome. One of the easiest ways to insure a successful outcome is to prepare a detailed written site-exploration program.

The site-exploration program should have several parts. The topics that should be addressed include site health and personnel safety; personnel requirements; equipment to be used; sampling methods to be used; and

TABLE 2.2. Outline of Site Remediation Plan.

Section	Contents
1	Introduction and objectives
2	Site description, location, and maps
3	Historical information
4	Surface soil information
5	Geological information—including state and local geological survey information and records
6	Site utilities—streets, sewers, overhead power lines, and buried water, gas, and telephone lines
7	Site maps and photographs—including aerial photos and zoning and property maps where appropriate
8	Site exploration plan—including alternative exploration techniques such as soil vapor probes, etc.
9	Sampling plan—how many and what kind of samples are to be collected
10	Analysis plan—how the samples are to be preserved, handled, analyzed, and stored, including the chain of custody documents for the samples, and laboratory analytical methods and sample holding times
11	Site safety and health plan—personnel safety and exposure information including personnel sign-off sheets

sample preservation and analytical techniques. The site-exploration program should also include a discussion of the exploration methods – both invasive and non-invasive – to be used, and the rationale for their use.

The site-exploration program documentation need not be unduly lengthy if standard procedures are already in place and have been submitted to and accepted by the site owner or operator. Table 2.2 suggests an outline for an initial site remediation effort.

The section on site health and safety should be part of the standard training program of the company performing the work, and the field manual should recap or highlight the exposures expected in the field and the initial safety precautions to meet anticipated personnel exposure levels.

2.4 INITIAL EXPLORATION

Site geology, site hydrogeology, and site chemistry must be addressed as the principal parts of any exploration and remediation program. The site geology is the foundation that supports the other investigations. As the foundation, it provides the support and the framework for the hydrogeology and chemistry investigations of the site. The hydrogeology of a site will describe the motion of water (and contaminants) in the site, to the site, and from the site. The chemistry of the site includes the information developed on water quality as well as the site mineralogy.

Some companies are using a standard set of drilling specifications which call for 5 (for example) test borings to be conducted to a specific depth. The drilling of three holes is necessary to determine the direction and flow of the groundwater, and to provide minimum site data. The information gained from the extra holes may or may not be sufficient to properly evaluate the site for the type of contaminants and the potential for cleanup. A more extensive initial investigation or a different type of investigation may be worthwhile.

One of the largest single cost items in an underground exploration program is the analysis of the samples collected. It costs very little more to collect many samples than it does to collect a few samples, if the samples are not analyzed. The optimum strategy for cost management is to collect a large number of samples, and put some into storage for later analysis if required. Some samples may degrade in storage, and others (such as water samples) cannot be stored. Soil samples can be stored if the storage conditions are adequate and the information used from these samples considers the age and possible decay of the sample.

2.4.1 SCS CLASSIFICATION SYSTEM

The purpose of the sampling program is to gain information. That information must be consistent across the site. It is often difficult to get two geologists to agree on the classification of a particular soil, partially because the classification system most geologists prefer is the subjective sand/silt/clay soil texture trilineal classification system, which consists of a triangular chart as shown in Figure 2.1. In this classification the boundaries between silt, sand, and clay are not well defined.

The trilineal classification system has gained wide acceptance. The classification system is adequate for agricultural uses, but provides little information about the properties of the soils. The system is also somewhat arbitrary, and does not account for the presence of larger sand and small gravel particles that may be present.

2.4.2 UNIFIED SOIL CLASSIFICATION SYSTEM

In order for geological observations to be uniform they must be supported by laboratory analyses. The Unified Soil Classification System is a uniform

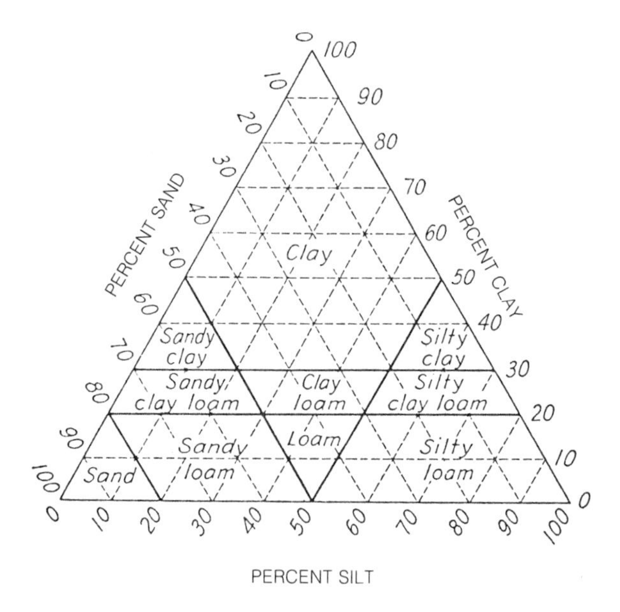

Figure 2.1 Soil Conservation Service classification system.

system of soil classification that is supported by low-cost physical testing for moisture content, Attaburg Limits (liquid limit and plasticity index) and screening analyses. The Unified System is also widely used by state highway departments, and by most geotechnical and foundation engineering firms.

The Unified Soil Classification System first divides soils into two classes based on grain size using a #200 sieve. Fine grained soils have over half the dry weight passing the #200 sieve. If more than 50% of the coarse grained materials are larger than a #4 sieve (4.75 millimeters) the soils are given the designation G (gravel), while they are designated as S (sand) if more than half are finer. The G or S is followed by a second letter indicating the type of gradation of the soils. The designation for these letters are P (poorly graded, uniform, or gap graded), M (contains silt), and C (contains clay or sand and clay). Fine grained soils are graded into three classes – M (for silts), C (for clays), and O (for organic mucks, silts, and clays). The fine soil designations are also separated by a second letter, which designates the relative moisture content of the soil and the difference between the moisture content of the soil when it behaves as a liquid and when it behaves as a plastic. The principal distinction between the fine grained soils is made on their ability to deform under load, or compressibility. The letter designations are H, which indicates a liquid limit over 50% (and high compressibility), and L, which indicates a liquid limit less than 50% (and low compressibility). The basis for determining which classification a soil fits into is the plasticity chart, which is shown in Figure 2.2.

The obvious advantage of this type of soil classification system is that it is uniform, and a sieve analysis made in a soils laboratory becomes the ultimate referee of what types of soil are present. A second advantage to the Unified Soil Classification System is that it can be related to engineering properties needed for cleanup of a contaminated soil. The disadvantage of the soil classification system is speed and cost. Laboratory analyses can take up to a week, and the cost per sample is approximately $50.00 for sieve analysis, and $30.00 for determination of the plasticity and liquid limits.

2.5 CHEMICAL ANALYSES

A good soil investigation includes both physical and chemical descriptions of the individual soil types and a hydrogeologic estimation of the behavior of the composite soil mass. The chemical information is the most costly to obtain, and the need for it must be balanced against the cost of obtaining it. When the scope of the investigation is limited, it would be better for the geologist to limit the chemical analyses to only the principal formations that may be contaminated, or those that may be directly involved in the transport or storage of the contamination.

The chemical tests generally include determination of specific items by a recommended ASTM or EPA or other test method. Laboratory testing may include the measurement of specific chemical properties such as ion ex-

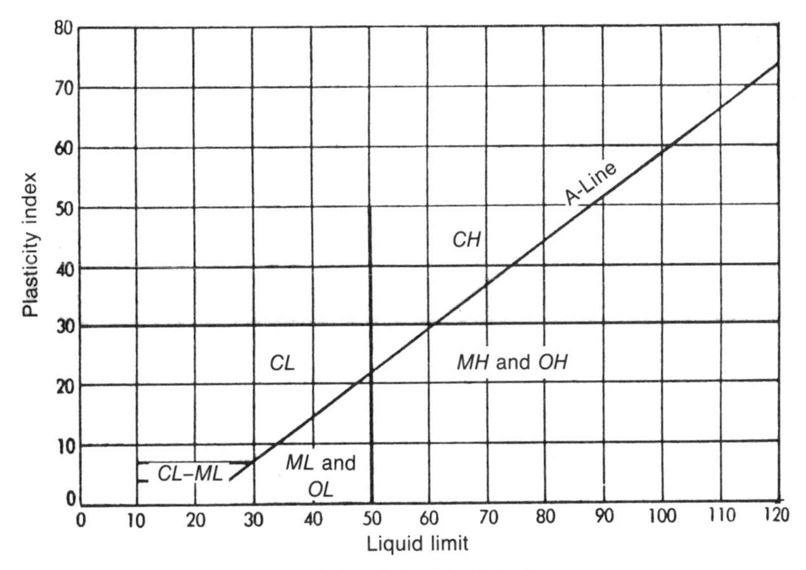

Figure 2.2 Soil plasticity limit chart.

change capacity and absorbance for a specific soil, as well as the determination of organic or inorganic soil constituents.

2.5.1 TOTAL PETROLEUM HYDROCARBON AND ORGANIC CARBON ANALYSES

The presence of petroleum in the soil can be detected by changes in the organic carbon measurement as well as by other tests. The organic carbon test by itself is not definitive because the range of contamination is often within the natural variability of the organic carbon content of many clayey and silty soils. In sandy soils, however, organic carbon may be a good indicator of petroleum contamination. The purpose of mentioning organic carbon analysis is that it is often less than half the cost of direct tests for petroleum hydrocarbons, and because it is an indicator of the partitioning coefficient of the soil, i.e., the ease with which the petroleum is retained or released by the soil.

Total petroleum hydrocarbons are commonly determined by several test methods. There is little consistency between the analyses required among the states. The most common analytical methods in use are Standard Methods #503, and EPA-SW-846 methods 9071 and 418.1. These methods are all of comparable sensitivity and reliability. The 418.1 method is more commonly used to detect the presence of the heavier fraction of diesel oils.

Petroleum hydrocarbons will be found in the unsaturated zone above the free product, or in migration pathways from the leak source. The quantity of hydrocarbons present is dependent upon the type of soil, water content, grain size, and other factors that include the quantity of organic material naturally present in the soil. It is estimated that gasoline concentration in the various types of soils is as shown in Table 2.3 [1].

2.5.2 SOIL MINERAL BALANCE

Each soil type is unique. It contains a set of chemical and size properties different from its neighbors. The chemical composition of a soil can affect its ease of cleanup. For example, either a highly organic soil, or one with a high clay content, will tend to adsorb hydrocarbons into the soil, while a

TABLE 2.3. Estimated Concentration of Distillates in Various Soils, Mg/Kg.

Material	Gasolines	Fuel Oils
Coarse gravel	950	4800
Coarse sand	2800	15,000
Fine sand and silt	7500	39,000

TABLE 2.4. Common Chemical Analyses for Soils.

Soil Cations	Soil Anions	Other Analyses
pH	Carbonates*	Soil exchange capacity
Calcium*	Phosphorus	Soil absorption capacity
Magnesium*	Chloride	Total organic carbon
Sodium	Nitrogen (various forms and total)	Total petroleum hydrocarbons
Manganese*	Silica*	Volatile organics
Iron*		Other hydrocarbons by GC or GS/MS methods
Potassium		BTEX by GC/MS

*These compounds are quite water soluble and can create solids deposition problems in strippers.

sandy soil may not adsorb gasoline as strongly because the adsorption sites are fewer and less polar due to the mineralogy of the soil. High levels of calcium in the soil have defeated *in-situ* bioremediation programs because the calcium reacted with the phosphate (added to the nutrient water to promote bacterial growth) and formed a crystalline plug around each of the wells, reducing the permeability of the wells by several orders of magnitude in just a few weeks of operation.

The principal soil cations and minerals that should be identified include calcium, iron, magnesium, potassium, sodium, and manganese. Principal inorganic anions that should be identified include nitrogen, phosphorus, sulfates, chlorides, and carbonates. Specific tests are available for each of these soil ingredients, and information on the chemical characteristics of the soil is necessary to conduct an effective remedial design.

The reason for the analysis of each of these minerals is to determine the principal ionic balance of the soil. The cations are water soluble, they can affect the ion exchange capacity of the soil water, and can create special needs or problems in the treatment system if the levels are excessive. The cations, especially calcium, magnesium, and iron have been known to cause problems in stripping towers. The anions, including phosphates and nitrogen forms, are all highly water soluble, and they can provide necessary nitrogen and phosphorus for supporting biochemical reactions.

Table 2.4 indicates the types of chemical, mineral, and laboratory analyses commonly run on soils.

It is often desirable to determine the pH and the Eh (Oxidation Reduction Potential) of the soil, especially in the contamination zones, as this is often an indicator of the potential success of a bioremediation effort should one prove necessary. The optimum pH of many soils is in the acid range, sometimes as low as pH 4.5. The optimum pH range for many of the bacteria that will degrade the petroleum is closer to a neutral pH range, generally between pH 6 and pH 8. The presence of negative Eh indicates reducing conditions; biochemical reactions can proceed under reducing, anaerobic, or oxygen-starved conditions, but the overall rate of microbial activity is between 2 and 30 times slower than under aerobic conditions.

Nitrogen, carbon, and phosphorus are key ingredients in supporting bacterial life. The carbon is generally in sufficient supply, as is the phosphorus. Several investigators have proposed the approximate C:N:P ratio of between 100:5:1 and 100:20:0.5 as necessary to support biological growth. The exact ratio may vary from site to site, but the key nutrients that are missing are often phosphorus and nitrogen. Nitrogen and oxygen for microbial respiration must often be supplied to allow the biodegradation to proceed rapidly.

2.5.3 WATER QUALITY

The site water quality data is a good indicator of the conditions beneath the site. Many site investigators are so wrapped up in looking for the principal contaminants that they do not sample for or do not analyze the major constituents in the soil. Analyses for organic contaminants or for benzene, toluene, ethylbenzene, and xylene (BTEX) alone are insufficient. The water chemistry must be addressed in a holistic fashion.

With the current emphasis on "regulated pollutants," and an incomplete understanding of the fundamentals of water treatment chemistry, many investigators fail to include analyses for the so-called "conventional" parameters in their analysis programs and focus instead on the trace organic contaminants. It is unfortunately quite common to see water quality analyses that fail to include pH and alkalinity analyses that are necessary for the engineer to design a water treatment system.

The analysis program as described in Table 2.5 is suggested as a minimum for determination of water quality parameters where gasoline or diesel contamination is suspected. The general analyses and the inorganic analyses should be performed each time a sample is collected and analyzed for *any* organic compound. Table 2.6 shows the recommended minimum sample volume, sample collection container, and holding time and sample volume required for the compounds listed in Table 2.5.

The information in the following table may at first glance appear excessive, and it might seem that a major sampling campaign will be required for the sampling and analysis of a single well. In reality, many of the sample bottles can be combined. A number of wells have been successfully sampled with the collection of as little as two to three gallons of water in less than six bottles.

The purpose of the analytical program is to develop a complete information package on the conditions present at the site. The cost of the general analyses (without the starred components) and the inorganic analyses is less

TABLE 2.5. Recommended Analyses for Water.

General Analyses	Inorganic Analyses	Organic Analyses
pH	Cations	*Benzene
Acidity or alkalinity	Sodium	*Ethyl benzene
Total solids	Magnesium	*Toluene
Filterable solids	Calcium	*Xylene
Volatile solids	Potassium	*Total petroleum Hydrocarbons
Dissolved oxygen	Manganese	Other analyses
Dissolved CO_2	Iron	
Dissolved methane	Ammonia	
Conductivity	**Other metals	
Temperature	Arsenic	
Ox/redox potential	Barium	
*Total organic carbon	Cadmium	
*Chemical oxygen demand	Chromium	
*Biochemical oxygen demand	Lead	
*Total organic halides	Mercury	
	Selenium	
	Silver	
	Anions	
	Chloride	
	Nitrate	
	Nitrite	
	Phosphate, total	
	Silicates	
	Sulfates	

*These analyses may also be required by regulation or by the assessment needs of the site.
**Additional analyses suggested in initial site survey for site evaluation.

than $200.00 per sample, or on a par with the cost of a single organic analysis.

After the initial range of water quality parameters have been established, Sims [3] indicated that in subsequent analyses, total organic carbon can be substituted as a water quality indicator rather than the more expensive GC/MS procedures or analyses for BTEX.

2.6 NON-WELL SUBSURFACE EXPLORATION TECHNIQUES

For many petroleum-contaminated sites, the contamination is relatively shallow, and the compounds in gasoline are relatively volatile. One popular subsurface technique is the use of soil gas and headspace analyses. The technique is used successfully with both a hand-held photo-ionization detector (PID) or a flame ionization detector (FID) or with a field gas chro-

TABLE 2.6. Recommended Sampling Volumes, Handling and Preservation for Water Samples [2].

Parameter	Required Sample Volume, ml	Preservation Method and Material	Holding Time (days)
pH	50	G,T, field, grab sample	1 hr
Conductivity	100	G,T, field, grab sample	1 hr
Temperature	1000	G,T, field, grab sample	None
Ox/redox potential	1000	G,T, field, grab sample	None
Dissolved gases	10 each	G,T, 4°C Dark	1 max
Tot. org. carbon	40	G,T, 4°C	1 max
Tot. org. halides	500	G,T, 4°C	1 max
Alkalinity/acidity	1000	G,T, 4°C	6 hr
Diss./susp. solids	1000 min	G,T, 4°C	1
Metal cations	1000 min	P,T, 4°C @ pH 2- HNO_3	180
Chloride	50	P,T, 4°C	7
Phosphate	50	P,T, 4°C	1
Silicates	50	P,T, 4°C	7
Volatile organics	50	G vial, 4°C	1*
Non metal anions	100-400 each	P,T,G, 4°C	1
Ammonia	400	P,T,G, 4°C @ pH 2-H_2SO_4	7
Organic analyses by GC/MS	4000	G, dark, 4°C	1*

*Analyses or extraction for later analyses must be made within 1 day.
P = polyvinyl chloride, G = borosilicate glass, T = Teflon.

matograph. The PID and FID are non-discriminating, and they cannot differentiate between different organic compounds such as alcohols, benzenes, alkanes, etc. Both the PID and the FID provide a simple yes or no answer to the question of whether an organic compound is present. Both are good overall tools for detecting the presence of organic substances and their approximate levels, but they are unable to determine which substance they are detecting. The PID is sensitive to non-methane hydrocarbon gases. The FID is calibrated with methane and will detect methane and all non-hydrogen gases. The PID is often calibrated with isobutane.

The field gas chromatograph, (GC) when used under the right field conditions and with proper calibration, can detect the relative differences in the principal compounds of gasoline, and can provide useful information about the relative direction and aging of the plume.

At a recent seminar on subsurface remediation held in January 1990 by the U.S.EPA in Atlanta and in other cities, a number of speakers suggested repeatedly that soil gas techniques be used to delineate the plume boundaries as an inexpensive alternative to drilling monitoring wells [4].

The principal difference between soil vapor surveys and headspace analyses is in the sampling technique. In a headspace analysis, a sample of the soil is collected and placed in a sealed glass container. The sample container is then submerged in hot water to heat the soil and volatilize the organic compounds. The headspace, the space between the top of the container and the top of the sample, is then sampled and analyzed for the presence of organic vapors, usually by gas chromatograph. The headspace analysis permits a limited number of replicate analyses to be performed on the sample, but the physical results cannot be duplicated once the analyses are complete. The procedure is somewhat more elaborate than soil vapor surveys, but it does provide for laboratory confirmation of the levels of contamination present at the time of sampling of the soil.

Soil vapor surveys have several advantages and a few disadvantages. The principal advantage is that it is quick and it provides a lot of information. The boundary of the plume can be established with a high degree of confidence, relatively quickly, without major disturbance at the site. A soil vapor survey, when conducted with a gas chromatograph, can delineate the difference between benzene, ethylbenzene, toluene, and xylene, and their degradation compounds. From the relative differences in the concentration of those constituents, a good analyst can predict the relative age and direction of the movement of the plume.

Since most soil vapor surveys are performed with a surface drill to open the pilot hole, and a driving tube, the practical range of the depth of the soil vapor survey is less than 20 feet. Soil vapor surveys can work very well in sandy and permeable materials, but the presence of a clay either in barrier layers or occluded lenses may prevent vapor movement, and can hide the soil vapor from the investigator.

One of the principal drawbacks of a soil vapor investigation is the fact that it is not reproducible. At the end of the survey, the pilot tubes are withdrawn, and the holes are plugged. The analysis for gases requires an experienced chemist, the gas chromatograph used in the survey requires frequent calibration, and great caution must be used in collection and handling of the soil gas to insure accurate measurements.

In sandy soils and soils low in clay, geophysical techniques can be used to locate subsurface contamination. Ground-penetrating radar, soil conductivity and soil resistivity surveys can, with some degree of accuracy be used to locate plume boundaries. All of the geophysical techniques are measurements of soil and soil-water conductivity. The presence of the dissolved hydrocarbons decreases the resistivity of the groundwater, which can be detected from the surface.

The presence of metal objects, overhead power lines, salt water, clays, or highly conductive formations (salt containing) can interfere with geophysical techniques, rendering them useless. The investigation technique must be suitable for the particular site conditions.

2.7 WELL-DRILLING TECHNIQUES

The selection of the best type of well and well-drilling equipment to use should be left up to the experienced geologist at the site. One of the most widely used drilling apparatuses is the hollow stem auger. With this tool the geologist can drill holes over 100 feet deep, and can install wells in the range of 2″ to 10″ in diameter. The auger can be used to install 2″ wells through the auger stem. The auger is fast and is suitable for many locations except where there is loose granular material such as sand or gravel, and then the sides of the hole will collapse. In those instances, the geologist will probably choose to use a drilling mud (natural or artificial) to hold open the hole while the excavation proceeds.

The Georgia Geologic Survey has provided some guidance on the selection of drilling rigs for the construction of monitoring wells at hazardous waste sites [5]. Tables 2.7 and 2.8 are reproduced from that source.

Table 2.7 compares the advantages and disadvantages of small-diameter wells for installation, monitoring, and pumping uses. Since this information was first presented in 1981, advances in well-drilling technology and in equipment for monitoring and pumping small-diameter wells have changed a few items; these items are denoted with an asterisk.

One of the major controversies regarding the drilling of wells is the use of drilling muds or fluids, and their effect on the development and chemistry of the wells. Anything introduced into a well as an aid to drilling will find its way into the formation.

The drilling fluids can modify the chemistry of the material surrounding the well and can influence the water quality and the permeability of the well. Geologists "develop" wells to attempt to remove the drilling fluids. Sometimes they are successful. One of the chief areas of concern is that the information gained about the water quality and the formation is after the fact, and that the act of installing the well creates changes that *may* lower the permeability of the formation or make other undesirable changes in the chemistry of the formation or the water immediately surrounding the well. It must be realized that the problems associated with the installation of a well are most acute at the beginning of the sampling period, which is precisely where many people are seeking to obtain initial data regarding the properties of the water, etc.

The question of how the well was installed and which drilling fluids were used seldom lasts beyond the first year.

Exploration or monitor wells should be 2″ in diameter. Wells to be used for development of the formation or pumping should be at least 4″ in diameter. The choice of whether or not to use a drilling fluid is best left to the driller, but should be discouraged unless absolutely necessary.

TABLE 2.7. **Guidelines for Drill Rig Selection.**

Anticipated Drilling Conditions for Well-Construction Program	Optimum Drill Rig	
	Auger	Rotary
1. Shallow water table (less than 20 feet)	Yes	Yes
2. Deep water table (greater than 30 feet)	No	Yes
3. Gravel and resistant zones	No	Yes
4. Loose sand or thick clays	Difficult	Yes
5. Undisturbed samples required	Difficult	Yes
6. Disturbed samples required	Yes	Yes
7. Depth to bedrock less than 80 feet	Yes	Yes
8. Depth to bedrock greater than 80 feet	Difficult	Yes
9. Rock coring	Yes	Maybe**
10. Seals and screens to be placed at specific intervals	Difficult	Yes

*Many rotary rigs are not capable of coring.
**Rotary drilling rigs often require drilling fluids.
The API Publication #1628 [6] summarized the basic well-drilling techniques, and their thoughts regarding the selection of a drilling rig are essentially the same as those presented below.

TABLE 2.8. **Guidelines for Selection of Monitoring Well Diameter.**

Anticipated Drilling Conditions for Well-Construction Program	Monitoring Well Diameter		
	2 inch	3 inch	4 inch
1. Shallow water table (less than 30 feet)	Yes	Yes	Yes
2. Deep water table (greater than 30 feet)	No	Possible	Yes
3. Well to be considered for decontamination or de-watering	No*	Possible	Yes
4. Well to be sampled frequently	Possible	Possible	Yes
5. Water level recorder to be installed	No**	Difficult	Yes
6. Clayey soils that may be difficult to develop [a well]	No	Possible	Yes
7. Bedrock coring NX-sized [2-1/8″ D.] or overburden to be cased off	No	No	Yes
8. Gamma logging anticipated	No	Yes	Yes

*New pump technology makes it possible to install a monitoring or production pump in a 2″ well. This practice is not recommended for any but very poor geologic formations with low well yields.
**New technology now makes it possible to install a water level in a 2″ diameter well.

2.7.1 CORE SAMPLES

The type of core sample that can be obtained is dependent upon the type of drilling equipment used, the geological formation, and the type of sampling equipment used to obtain the core sample. In many instances where the soil is moderately soft, the Shelby tube sampler is the preferred sampling method. The Shelby tube sampler is a thin-walled (16 gauge seamless steel) cylinder that is advanced by pushing it ahead of the drill bit or boring tool. The Shelby tube sampler is generally preferred because it disturbs the soil the least, but in unconsolidated material (sand and gravel) it is difficult to collect a sample unless a spring attachment is used to keep the sample in the sampling tube. Shelby tube samplers come in several sizes, but the most popular are the 2″ and the 3″.

Perhaps the second most popular type of sampler in use is the split spoon sampler. This sampler contains a driver head, and is driven into the soil by dropping a 140 lb. hammer twelve inches. Physically the split spoon sampler is 2″ O.D., and 1-3/8″ I.D.; the tube for the sample collection is between 18″ and 24″ long. While this is also a good sampler, the driving head causes the wall to be thicker, and it disturbs a sample slightly more than the Shelby tube sampler. The principal use of the split spoon sampler is in geotechnical and foundation work where the number of hammer blows required to drive the sampling head 6″ into the ground has been related to the load-bearing capacity of the soil. Either sampler will work adequately for subsurface contamination investigations, but neither can effectively be used to sample hard formations, including rocks and subterranean boulders and cobbles larger in diameter than about 3″.

In certain circumstances it is almost impossible to obtain a core sample from a formation. At the Hanford Nuclear Test Center for example, some of the soils are so hard that the blow count for a standard split spoon sampler has occasionally exceeded 1100. In order to overcome soil resistance a much heavier driving hammer is used. This driving hammer weighs over 400 pounds.

In order to drill into the rock, the driller must use a special drill bit. Most of these bits are capable of collecting core samples up to about 4″ in diameter, but one of the most common sizes is the NX bit, which has an outside diameter of 2-11/16″, and produces a core sample of 2-1/8″ diameter. The popularity of the NX bit may be related to the drilling of 2″ monitor wells in rock.

The drilling in rock must be far enough to determine the fracturing of the rock. In many instances, the top of the rock is weathered, split, or decaying. In limestone, carbonate, and other formations, the rock may have substantial cracks or joints, and/or solution cavities. In sandstone formations, the rock may be lightly cemented, and may be permeable. Many subsurface exploration contracts call for drilling a set distance into unbroken rock such as 10 or 15 feet; the depth of rock penetration should be dependent upon the likelihood of cracking or cavities, and the permeability of the rock itself. The initial remedial investigation, performed before the drilling is begun, can provide some useful information and guidance to the geologist and the drill rig operator.

Some state codes require monitoring of the major aquifers and the rock. In order to drill into the rock, the driller must use a special technique to insure that contamination from the formation above does not enter the rock through or around the well. In order to do this, the well driller must first drill a hole down to the rock, and penetrate a short distance into it. This hole is then cased for the entire length, and grouted. During the grouting of the well the casing is pulled up slightly so that the grout is forced out around the bottom of the casing to insure an adequate seal between the casing and the formation. A new, smaller-diameter well will then be drilled inside the casing through the bottom of the well plug into the rock. This is commonly referred to as a Type III well, which will be discussed below.

A drilling program should collect core samples at specific distances as the investigation proceeds. Samples should be collected by Shelby tube or by split spoon sampler at least every 5 feet in non-rock materials, or more frequently if the formation changes. In rock, a continuous core sample should be taken. When drilling into rock, the exploration should penetrate the formation at least 10 feet and preferably 15 feet into the unweathered material.

2.8 HYDROGEOLOGIC INFORMATION

2.8.1 SLUG TESTS

The slug test is one of the most common and least expensive ways of obtaining information about the permeability of the underground formations. Once the well has been installed and developed, and the well screen put into place, one or more slug tests should be performed.

There are two types of slug tests: slug in and slug out. In the "slug in" test, a solid plug of known dimensions is suddenly lowered into the well, displacing the water; alternatively, a volume of water is poured into the well, raising the level. The slug displaces the water in the well, raising the water level. The increase in water level creates a hydraulic gradient that forces water from the well back into the formation. The water level surface elevation is carefully monitored with a tape or with a DataLogger, and by the procedures established by Bower and Rice [7,8] for fully or partially penetrating wells in confined and unconfined aquifers, the falling head is related to the hydraulic conductivity of the formation. In the "slug out" test, a volume of water is withdrawn either by bailer or by removing the plug used in the "slug in" test, and the water level is again measured.

Care must be taken in use of the slug test because it is of short duration and small volume. The slug test has been known to accurately measure the permeability of the sand pack around the well, especially in highly conductive formations, in wells with large screened intervals, and in wells with small diameter [9]. Multiple slug tests, at least one "slug in" and one "slug out" test, should be run at each screened interval and with each well.

Slug test data interpretation is somewhat subjective. The estimate of the well capacity should be based upon the linear portion of the slug test, preferably at or near the end of the test. According to Bower, this gives results that are the most reproducible and which conform most nearly to the actual values that will be encountered when the wells are pumped. Slug test data from 2″ diameter monitor wells should be used only sparingly, if at all, and only to develop preliminary data from piezometric nests of wells (wells closely spaced but at different depths).

2.8.2 PUMPING TESTS

The long-term performance of a well is not easily estimated from a short-term duration test such as a slug test. Where multiple wells are to be installed, and there is to be extensive manipulation of the groundwater regime, one or more pumping tests should be conducted. One of the strongest reasons to perform a pumping test is to aid in the selection of the pump, to match pump performance to aquifer performance. Two principal consequences of mismatching the pump and the aquifer performance are: (1) burnout of the well pump, which will require frequent pump replacement; and (2) delayed cleanup of the site, or possible spreading of the contaminant plume because the projected hydraulic gradients were less than anticipated due to low pumping rates.

At least one pumping test should be run per site. Depending upon the formations being pumped, a pumping test of between 8 and 96 hours duration should be conducted to adequately stress the aquifer. The pump test, once started, runs continuously until all data is gathered. In some states where unions are strong, the site may have to have an operating engineer on site (three per 24-hour day) to monitor the performance of the equipment for the duration of the test. Provision also must be made with the regulatory community for the treatment of the pumped waters and disposal of the contaminants removed from them.

The pump test will determine well drawdown at the given pumping rate. A well should be supported with adequate piezometers to insure that sufficient data is obtained from the pumping test. From the drawdown in the piezometers around the pumping well, and from water data in the pumping well, the true yield of the well can be calculated. During the pumping test, frequent water quality samples should be taken to determine whether or not the water quality will change over the duration of the pumping test. The monitoring requirements for water quality during the pumping test need not be extensive. Periodic monitoring for conductivity, temperature, CO_2, and total organic carbon and/or total petroleum hydrocarbons may be all that is required for adequate evaluation. At or near the end of the pump test a more extensive determination of the water quality should be made.

If more than one formation of different types and depths is to be pumped, it may be warranted to conduct more than one pumping test. Where long-term recovery of a dissolved phase relies upon a recovery pump and a depression pump, a pump test will help develop necessary engineering information about the long-term performance of each pumping system and the overall yields of the product pump and of the well pump.

In a moderately permeable formation with a recovery pump for free product, and a depression pump, it is reasonable to assume that there is approximately $4000 to $6000 invested in the pumps, controls, and equipment per well. If either pump burns out, the replacement cost will be higher because one or both pumps must be pulled, the new equipment installed, and the controls reset.

There are a number of ways in which both centrifugal and air-driven pumps can fail. One way for a centrifugal pump to fail is to overheat the motor. Centrifugal pumps operate within a relatively narrow range of head and flow conditions. If the system head is too high and the flow is too low, if the system head is too low and the throughput is too high, or if the well does not provide sufficient water to cool the pump, the pump motor may be overloaded and fail.

While air-driven pumps are somewhat less sensitive to the same type of mechanical failure as the centrifugal pump, the air-driven pump capacity is also limited. Typical design specifications require about 1 cubic foot per minute at 75 psig per gallon per minute pumped. If the compressor air tank is too small, or if the well capacity is too great, the air compressor tank will be cycle rapidly, as the demand for high-pressure air may temporarily exceed the supply. Rapid cycling of the air compressor motor can be just as bad as rapid cycling of the pump motor, and will burn it out just as quickly. If the pump capacity is too small for the aquifer, the cleanup will be delayed, and the desired water depression levels will not be obtained.

The data obtained from a pumping test will permit the designer to be more realistic in determining the long-term performance of the treatment equipment, and to be more accurate in determining the size of the equipment. At many sites the initial data presented to the designer is so sparse that the designer is forced to make extremely conservative assumptions regarding the equipment size and performance. The conservative nature of these assumptions may cause the treatment system to fail.

2.8.3 SOIL VAPOR STRIPPING

Just as the hydrogeologist should take the extra time and effort to develop a body of information about the hydraulic characteristics of the site and to obtain data on the long-term pumping rate for the liquid stream, he or she should also be equally cautious in developing long-term soil venting data for the remediation system. According to current literature [10-12] soil venting tests are generally run for a limited time to gather vacuum data. Most venting tests are run for a period of 2-4 hours at each well. If air samples are taken during that period, it is usually only one sample. To make vacuum readings, the test points are generally set out in one or two lines from the well, and the test probes are driven between 5 and 15 feet into the soil. Any calculations are based upon the assumption of a uniform cylinder of soil. At many sites the soil has already been shown to be non-homogeneous, and the discontinuities in the soil need to be addressed by additional measurements. The analysis of the soil does not address one problem generally encountered at petroleum retail stations—the fact that the soil is covered on top by gravel, and over that by a layer of concrete. Information in the report by Mutch et al. [13] indicates that the air may move differently in each of the portions of the soil in response to the permeability of the formation.

At a typical service station site, the pavement will act as a large barrier to air movement, and will prevent air from easily entering the soil in the vicinity of the extraction well. As a consequence, the air entering the soil may come from an area much nearer the edge of the pavement, and will vapor-strip an area much larger than anticipated. When rainfall saturates the soil adjacent to the paved areas, it has been observed that the airflow in the soil venting system decreases, but it is not clear where the soil vapor is coming from.

Soil vapor stripping will de-water the soil. Most soil vapor extraction studies cannot account for this effect. Over a longer term, the dehydration of the soil is detrimental to the extraction program because the loss of water often causes the fuel to cling to the soil, decreasing its removal rate; lack of adequate soil moisture can also adversely affect a bioremediation effort.

Soil venting tests should be conducted at each well for a period of several hours or until equilibrium vacuum conditions are attained at the outermost probe at the shallowest elevations. The tests should be conducted using a vacuum pump capable of producing at least 5″ mercury gauge (approximately 65″ water column) or more. At or near the end of the test, a number of test lines (at least 2 and preferably 3 or 4) should be drilled at each prospective vapor recovery well, at radial distances that proscribe equal volume cylinders. Vacuum readings should be taken at distances just below the slab, and at periodic intervals down to the top of the well screen, or as deep as is practical. The instruments used in obtaining the soil vacuum levels should have a minimum sensitivity of 0.01″ of water. The soil in the vicinity of the well should be tested for water content, and the geologist should take care to attempt to estimate the height of the water and the hydrocarbons in the vadose zone surrounding the extraction well.

2.9 BIOREMEDIATION DATA REQUIREMENTS

Microbial degradation of chemicals in soils is a subject which, while it is becoming more widespread in cleaning up soils, is still poorly understood [14]. The specific mechanisms and degradation pathways are being investigated, but the research is still in its infancy. Just a few years ago, for example, it was believed that all soil below a depth of about 6 feet was sterile. Current research activity is underway determining which types of organisms live in the "deep" soils, and how they degrade the food available to them.

If there is microbial activity taking place it will either be aerobic or anaerobic. The end products of aerobic biological activity are carbon dioxide and water. The principal end product of anaerobic activity is methane. Both of these gases can be measured in the soil, and in the water. Another indicator of aerobic activity is the presence of dissolved oxygen in the groundwater.

While it may not be necessary to determine the type of organisms present in the ground, it is necessary to determine what their overall chemical composition is, and how to optimize it. Carbon, nitrogen, and phosphorus in the soil column appear to be the three key indicators of the viability of biological activity. Analysis of these substances has been discussed previously, and is recommended for inclusion in the analysis during the initial exploration of the site.

If it becomes necessary to determine which types of organisms are present, a microbiologist can identify the soil microbes from a sample of soil, provided that sample has not been sterilized by exposure to high temperatures, or permitted to dehydrate. The practice of storing the soil samples at or below room temperature in sealed glass containers will generally permit recovery of the soil microbes if the storage time does not exceed several months.

While it may be desirable to perform some limited biological viability tests on the microorganisms present in the soils at the contamination site, there is little need for an elaborate pilot program unless the owner or operator needs the reassurance that the system will work, or needs to know the length of time required for cleanup. With regard to the latter point, it is often pointless to determine the rate of removal in the laboratory because field conditions may be substantially different, and the owner or operator will not be allowed to discontinue cleanup operations at the site until it meets state criteria, no matter how long that takes.

2.10 DISPOSAL AND INCINERATOR DATA REQUIREMENTS

When the volume of contaminated soil is small, and the depth of the contamination is less than 20 feet, the owner or operator may choose to remove the soil and dispose of it in a sanitary landfill, or a hazardous waste landfill. If the volume of soil is large enough, disposal by incineration may also be a commercial possibility. South Carolina, Virginia, Tennessee and many other states encourage this type of disposal.

The data required for disposal of the soil in a landfill will vary from state to state. In many states, petroleum-contaminated soil is not a hazardous waste, and under U.S.EPA criteria, there is a specific exclusion from the requirements to run the Toxic Characteristic Leaching Procedure analyses used to determine whether or not a waste is a hazardous waste (40CFR261) for product petroleum-contaminated soil in other states, however, the TCLP analysis is required for all soil to be sent for disposal. In the event that a soil is being analyzed for incineration disposal, total petroleum hydrocarbons, BTEX, and lead analyses should be run on composite samples of the soil to satisfy the incinerator and the local authorities.

If incineration is a disposal option to be explored, some additional information should be obtained in order to better evaluate the viability of the option and the cost of disposal. Data required for incineration evaluation should include grain size distribution, moisture content, metals content, volatile organics, total organic carbon, sulfur, total organic halides, ash content, BTU content. If the soil is to be destroyed in a high-temperature combustor (if any of the components of the soil have a volatilization temperature above 1500°F) analysis of the soil and ash for metals, ash softening, melting, and fluidization temperatures should also be performed.

For most soils, a primary incineration chamber of less than 800°F is generally used, and at that temperature, the soil is well below its melting point. Incinerators are extremely unforgiving if the soil fuses into a liquid plug. If ash and melting analyses are indicated, between three and ten representative samples of soil should be analyzed and fired, depending upon the size of the site. Grain size analyses need only be sufficient to characterize the range of soil sizes encountered, preferably the minimum and

TABLE 2.9. Soil Incineration Data Requirement.

Information Required	
General	**Soil Information**
Volume to be incinerated	Grain size distribution
Depth of excavation	Sulfur content
Depth of water table	Chloride content
Map of site, and local areas	Range of volatile organics in soil
Utilities available	Heating value for soil
Permitting authorities—names and addresses	Ash fusing, softening and melting temperatures
Terms of contract	Total organic carbon
Applicable state and local air, water and solid wastes standards	Metals in soil
Disposal Analysis	Antimony
Analysis of ash for RCRA metals	Arsenic
Final disposition of incinerated materials	Barium
Handling and disposition of scrubber water and scrubber sludge	Beryllium
Handling and disposition of activated carbon if required for scrubber water	Cadmium
	Chromium
	Lead
	Mercury
	Silver
	Thallium

maximum distributions. Tests for BTU content, TPH, ash analyses, sulfur, and total organic carbon should be performed frequently enough to determine the upper and lower ranges, and the statistical variation.

In the event that the soil is classified as a RCRA hazardous waste, additional metals for analysis will include chromium, arsenic, thallium, silver, mercury, lead, beryllium, barium, cadmium, and antimony. Despite the fact that many of these metals may occur naturally in the soil, the EPA has established discharge limits for air emissions from incinerators where the background concentration in the feed exceeds certain criteria levels [15]. To provide sufficient information for the incinerator contractor, the presence of these metals in the soil must be considered. The information required by an incineration contractor is summarized in Table 2.9.

2.11 SUMMARY OF DATA REQUIREMENTS

Each type of remediation effort will have different data requirements. The items of information important to one type of effort may not be required for a different type of effort. Table 2.10 outlines much of the information discussed previously, and displays the type of information that will be common to all remediation efforts and investigations.

TABLE 2.10. Summary of Soil Remediation Data, Information, and Regulation Requirements.

Type of Investigation or Remediation Effort	Regulations						Data Type										
	Water	Sewer	RCRA	Air	Waste	Other	Geology	Water Table	Soil Size	Boring Logs	Hydrogeology	Water Chem	Soil Chem	Vol Organics	Soil Metals	Pump Tests	Combust Analysis
General			X		X	X	X	X	X	X	X	X	X	X	X	X	
Excavate/remove			X		X	X	X	X	X	X	X		X	X	X		
Pump-and-treat	X	X		X		X	X	X	X	X	X	X		X		X	
Vapor stripping				X	X	X	X	X	X	X	X			X			
Bioremediation	X				X	X	X	X	X	X	X	X	X	X	X	X	
Incineration	X	X	X	X	X	X	X	X	X	X	X			X	X		X

REFERENCES

1. American Petroleum Institute. 1989. *A Guide to the Assessment and Remediation of Underground Petroleum Releases, 2nd Edition*, Publication 1628.
2. U.S.EPA. 1987. *Groundwater Handbook.* Ada, OK: R.S. Kerr Center. Publication Number EPA/625/6-87/016.
3. U.S.EPA. 1989. *Seminar on Site Characterization for Subsurface Remediations.* Technology Transfer, Sept. 1989, Publication Number CERI-89-224.
4. *Seminar on Site Characterization for Subsurface Remediations*, op. cit.
5. McLedmore, William H. 1981. *Monitoring Well Construction for Hazardous Waste Sites in Georgia, Circular #5.* Georgia Geologic Survey.
6. op. cit.
7. Bower, H. and R. C. Rice. 1976. "A Slug Test for Determining hydraulic Conductivity of Unconfined Aquifers with Completely or Partially Penetrating Wells," *Water Resources Research*, 12:423-428.
8. Bower, H. "The Bower and Rice Slug Test—An Update," private communication.
9. Bower, H., op. cit.
10. EA Engineering. Internal memorandum on soil venting, 1989.
11. Sims, R. 1990. *U.S.EPA Seminar on Site Characterization for Subsurface Remediations, Atlanta, Georgia, January 16, 17, 1990.* EPA Publication Number CERI 89-224.
12. Mutch, R., A. Clarke, J. Clarke and D. Wilson. 1989. "*In Situ* Vapor Stripping: Preliminary Results of a Field-Scale U.S.EPA/Industry-Funded Research Project," *Annual Conference Presentation of the Water Pollution Control Federation, October 1989, San Francisco, CA.*
13. Mutch, R., A. Clarke, J. Clarke, and D. Wilson, op. cit.
14. U.S.EPA. 1990. *Abstracts of Speakers' Papers, "U.S.EPA's Biosystems Technology Development Program," February 13-15, 1990, Arlington, Virginia.*
15. U.S.EPA. 1989. *Handbook on Setting Permit Conditions and Reporting Trial Burn Results, Volume 2.* Cincinnati: Risk Reduction and Engineering Laboratory, Center for Environmental Research. Document Number EPA/625/6-89/019.

CHAPTER 3

Remedial Options

3.1 INTRODUCTION

THE selection of the best remedial option for a particular site should be made only after the contamination has been completely identified and described. The technology selection process must consider the type of contamination present, site-specific geology, chemical and physical properties of the contamination, the areal extent and depth of the contamination, the effectiveness of the technology in achieving the desired treatment levels, and the total cost of the project.

For petroleum-contaminated soils, the physical properties of the gasoline or diesel can be used to assist in the remediation effort. Density, volatility, solubility, viscosity, and biodegradability can all be used to advantage if the cleanup and treatment system is carefully selected. No one specific remedy represents the optimal selection for all sites. At each site, the design engineer must evaluate the possibility of using a specific remedy and then must attempt to evaluate the cost of that remedy for the site.

Table 3.1 presents some of the many options available for remediation of gasoline-contaminated sites. Each of the technologies outlined in the table will be discussed in greater detail in the following pages. Of the possible technologies available [1], only those most appropriate to petroleum remediation have been selected. Of the many new technologies available for considered, the only one selected for consideration was vitrification because it appears economically viable in certain situations.

3.2 ASSOCIATED PROBLEMS AND CHALLENGES

It is inevitable that remedial activity will cause some disruption of the normal routine at any facility, since drilling, pavement removal and replacement, etc., are all a part of the process of investigation and equipment

TABLE 3.1. Remedial Options for Petroleum-Contaminated Sites.

Remedial Option	Principal Use, Advantage	Disadvantages
Excavation and disposal	Shallow contamination, depths of less than 20 feet. One-time removal of soil.	High disposal and transportation costs, availability of disposal sites, long-term liability as a potential waste contributor to a landfill. Does not address removal or treatment of contaminated groundwater.
Drainage trenches and galleries	Can be used at depths of up to 25′ or more; provides positive removal of groundwater and free product. Can be used to provide infiltration control.	Does not remove large volumes of contaminated soil; collected liquids must be pumped to a treatment system. Trench bracing may be required for depths over 5 feet.
Wells	Can be used at the greatest depths. Will collect water and free product. Most commonly used systems. Equipment widely available and relatively inexpensive. Good in formations where water is at great depth and other solutions will not work.	Requires air or electric lines to be run to each well; controls for wells are moderately expensive; seasonal changes in water levels may require resetting of pumps. Collection of separate phases of petroleum and contaminated water is quite difficult when accomplished down the well.
Soil venting	Highly effective in removing volatile component from soil. Relatively inexpensive to install and operate. Quickly removes volatiles.	Will not remove semi-volatiles. Only removes about 50% of petroleum products. Should be used in connection with other technologies. Venting may require incineration for air pollution control.
Incineration	Complete disposal and cleanup of soil. Produces sterile ash suited for all backfilling. Remedial solution completed fairly quickly once permits are obtained.	Requires excavation of soil; will require air permits; will concentrate many metals in the ash; does not address existing groundwater contamination. Incineration has a high first cost, and may initially appear more expensive than other solutions.

TABLE 3.1. (continued).

Remedial Option	Principal Use, Advantage	Disadvantages
Bioremediation	Provides complete treatment without excavation. Can decontaminate difficult-to-reach areas. One of the least expensive technologies for long-term site cleanup.	Requires initial treatability study. May cause bio-fouling of wells. Technology may be difficult to manage.
Containment	Used to immobilize contaminants or control groundwater movements. Not a treatment system.	Does nothing to degrade or remove contamination. Only stops fluid migration.
Solidification	Used to immobilize contaminants and prevent leaching. Primarily used with a cement formulation.	Present solidification technologies are not generally useful on organics. Volume and weight increase of around 30% due to additives.
Vitrification	Relatively recent technology. Appears promising but expensive. Converts entire mass of soil to a glass block.	No long-term data on effectiveness of technology to control leaching from end product. Initial cost estimates indicate high cost. Maximum depth limited to about 20 feet. Heavy power demand. Will require air pollution control permits.

installation. The disruption can be minimized by selection of the appropriate technology and by careful planning and control of the contractor's work and schedule. Ideally, if the investigation and installation of the remediation can be scheduled around the reconstruction of a service station, disruption at that site will be minimized.

Many remedial systems will require years of operation before they achieve acceptable levels in the soil or groundwater. A ranking of the solutions from the fastest to the slowest is shown in Table 3.2.

One of the greatest challenges to the remediation engineer and the site owner is the ultimate problem of finding the most cost-effective solution. Cost elements that should be considered include the operating cost, as well as the cost of chemicals, exploration, analyses, and maintenance manpower.

Each technology has some type of ancillary equipment. The equipment may be simple or it may be complex. The equipment has an associated operating requirement that is often overlooked when budgets are prepared.

TABLE 3.2. Duration of Remedial Activities.

Activity	Duration
Excavation	Days to weeks. Location of an acceptable disposal site and securing necessary tests and approvals may require weeks.
Incineration	Weeks after permits obtained. Permitting process may require months.
Soil venting	Weeks to months for partial cleanup. Cleanup is rapid for volatile components only.
Bioremediation	May require months to years of periodic monitoring depending upon extent of contamination and rate of biodegradation and levels to be obtained.
Pump-and-treat systems (including trenches and drain, and well systems)	Remediation is generally quite slow depending upon the movement of water through the ground. Without treatment or removal of contaminated soils, treatment may require many months to years of operation before closure levels are reached.
Containment and vitrification	Since solutions are primarily used for containment they do not provide treatment and very long-term (10 years or more) monitoring may be required.

The removal of petroleum from an underground location will create a by-product or an effluent stream requiring further treatment. Each technology has its own unique by-product, and some will produce streams subject to regulatory purview. The disposal of the by-product streams must be considered when the total project costs are calculated. For example, carbon adsorption systems have spent carbon that must be regenerated. The spent carbon is often classified as a hazardous waste, and disposal is expensive.

Obtaining permits for some of the by-product and effluent streams may be more of a challenge than the installation of the technology. A stripping tower may require an air permit for construction and another permit for operation. The effluent stream from the tower may need to be passed through a demister to reduce the water droplets in the stream and then through a vapor combustor to destroy the benzene and other hydrocarbons. The process of obtaining the permits for the stripper and its ancillary equipment may be one of the most difficult aspects of the remediation process.

Similarly, the reinjection or introduction of any water to a site that does not meet drinking water standards will likely be regulated by a state's Underground Injection Control Program. Under this program, the remedial activity may require an RCRA permit or its equivalent. These permits are not easy to obtain.

3.3 EXCAVATION

As a remedial option, excavation is unsophisticated. Depending upon the depth and extent of contamination, the only requirement concerns earth-moving equipment and trucks. Pick it up and put it somewhere else.

Under the revised hazardous waste rules and regulations in 40CFR 261, petroleum-contaminated soil is considered a hazardous waste after 1992. Waste soil for land disposal must be accompanied by an analysis using the Toxic Characteristic Leaching Procedure. If the regulatory concentrations are determined to be above the TCLP limits, the soil will be a hazardous waste. The disposal of gasoline-contaminated soil in a lined, non-hazardous, or in an unlined landfill is being permitted on a case-by-case basis in some states.

Some states will permit or require aeration and landfarming of the contaminated soil before it is landfilled. Depending upon the total petroleum hydrocarbon or BTEX levels in the soil, the states may require treatment or may permit direct disposal. There is no general guideline and no consistency between the states.

3.3.1 DEPTHS OF EXCAVATION

All excavations over 5 feet deep into which workers may enter must have shoring, or they must be cut back so that the workers will not be buried if the sidewalls collapse. (This is an OSHA requirement.) Many backhoes can reach between 12 and 15 feet below the footing of the tractor, and this serves as a practical limit to inexpensive excavation. Deeper excavations are possible with extensive shoring and bracing, but may be prohibitively costly.

3.3.2 NUISANCE PROBLEMS

Dust, noise, and truck traffic problems are often particularly acute at an excavation site. These problems can be controlled. The contractors frequently allow the general public and neighbors to suffer the inconvenience because it costs a little more to do it correctly. Many of the problems generally associated with construction can be avoided by good contract management and selection of contractors who are sensitive to public relations problems and the owner's image.

3.3.3 GROUNDWATER

The groundwater table serves as a practical limit on the depth of an excavation. Extensive excavation below the groundwater table will require de-watering. Without de-watering, the excavation bottom may become a thin mud, which is very difficult to excavate and which must be de-watered before it can be treated or sent off for disposal. Depth to groundwater varies with each site. Any generalization about the presence or depth of groundwater at a specific location would be speculation unless one was familiar with the regional geology.

If groundwater is encountered in the excavation it must be collected and treated, as it will be contaminated with whatever is in the soil. The collection and treatment of groundwater can be costly because it will frequently contain silts and clays as well as the organic contaminants at the site. The groundwater must be pumped out of the excavation at a rate that will permit work to continue. On a recent Superfund project in New Jersey which involved excavation and incineration of the soil, the treatment of the groundwater was approximately 25% of the total project cost, or approximately $4 million of a total project cost of $16 million. The water treatment equipment was not elaborate, and treatment technology employed was not the most expensive option available.

3.3.4 ROCK EXCAVATION

The presence of rock and boulders is a perpetual challenge to any excavation-based remediation effort. Rock is not all alike. Weathered rock, limestone, shale, granite, gneiss, and sandstone all behave differently. Some are soft and porous while others are hard and brittle.

It is more expensive to excavate rock than soil. Frequently, rock excavation requires special construction equipment and/or blasting. For example, rock that cannot be removed by a dragline can be removed by a backhoe without blasting.

Rock blasting requires special precautions and special equipment and permits. Blasting is best left to specialists in the construction field, not to the general contractor. When optimal blasting techniques are used, rock comes apart in large boulders or chunks. When the rock is removed from the excavation, it must be decontaminated and/or sent off for disposal. The decontamination of the rock may be surficial if the rock is non-porous. If the rock is porous, decontamination may be very difficult. Either way, when the excavation encounters rock, the cost of the remediation will increase.

3.3.5 BACKFILLING

Backfilling an excavation is not as easy as it may first appear. The principal concern is the structural stability (load-bearing capacity) of the excavation and the settling of the excavation after it is backfilled. Unless the backfill is properly compacted, excessive settling of structure foundations may result. In the instance of a service station, this could result in the settling of pavements, the development of cracks in retail structures, and the settling of underground tanks and lines, causing their resulting rupture.

The proper backfilling and tamping of an excavation requires a geotechnical or civil engineer familiar with backfilling and compaction. Typically, soils are compacted to about 95% of their maximum density, and the backfilling is carefully supervised by the engineer and frequently tested by the laboratory. Proper backfilling also requires careful preparation of specifications that control the contractor's activities and payments. The specifications need not be overly complex, but an owner's representative should be on-site to insure that the work is being performed properly.

3.4 TRENCHES AND DRAINS

Trenches and drains are an old technology for control of surface and underground water. The advent of perforated plastic pipe, decreased cost of installation, a better understanding of the movement of groundwater, and the use of barrier technology has made the use of trenches and drains more widespread in remediation technology. Trenches and drains are used as collection and dispersion points for groundwater.

Figure 3.1 shows a typical layout of a drainage system. Note that the water table is lowered in the vicinity of the drain and the drainage lateral lines.

Trenches are large elongated drains packed with gravel. Infiltration galleries are drains that operate in reverse, delivering water or nutrients to the ground. Drains are multi-directional in that water will flow into them from all directions – top, sides, and bottom. With the advent of plastic liner technology, one side of a drain can be easily lined, making it less permeable but not impermeable. This is shown in Figure 3.2.

The drain shown in Figure 3.2 is shown only in profile. The side with the membrane (water stop) is keyed into an aquaclude (barrier layer). Water will still be drawn into the drain from around the ends of the barrier. Drains are not just one-directional.

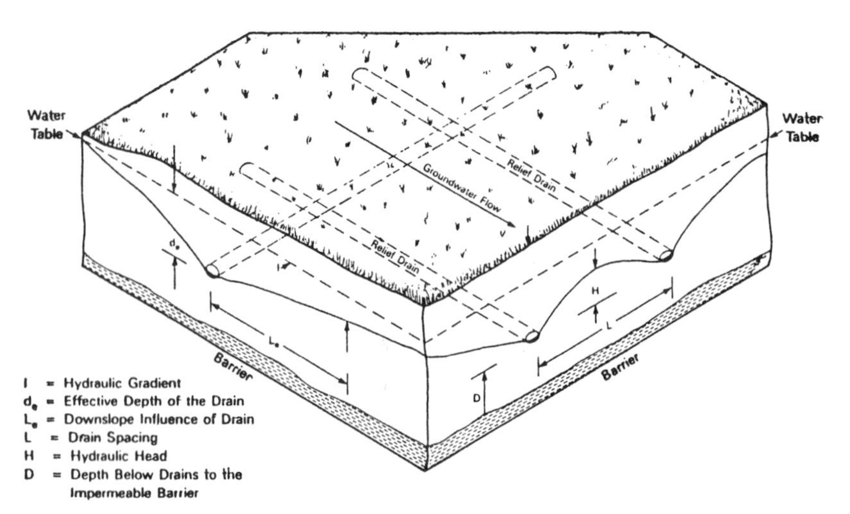

Figure 3.1 Typical layout of a subsurface drainage system. Source: Soil Conservation Service–1973.

Water moves through the ground very slowly. The general water velocity is measured in feet per day for very permeable formations to feet per year for tightly packed clays. One general measure of the velocity of water is the permeability of the formation. Permeability is measured in units of velocity–centimeters per second, or feet per day in English units. A well-drained gravel may have a permeability of 10^{-2} to 10^{-3} centimeters/second. A relatively impermeable clay will have a permeability of 10^{-7} centimeters/second. English units for permeability are in feet per day; the 10^{-2} cm/second is equivalent to 28.3 feet per day. The permeability of the tight clay (10^{-7} cm/second) is equivalent to 0.1035 feet per year. The permeability of the formation is only a measure of the ease with which water flows through. Permeability and velocity of the water through it are not the same thing.

The governing equation for the movement of groundwater through the soil is:

$$V = k * i$$

where

V = velocity
k = permeability of the formation (has the units of velocity)
i = hydraulic gradient (dimensionless–expressed as a decimal)

To find the specific rate of movement through a formation, one must divide the velocity by the porosity of the formation. For example, if the velocity is determined to be 40 feet per day and the porosity of the formation is 0.25, the actual movement of the pollution through the formation would be around 160 feet per day. Well drillers often express the yield of a well in gallons per day. This is also a unit of velocity, although the units are unconventional.

Water moves only in response to a difference in hydraulic head or hydraulic gradient. Hydraulic gradient is the slope of the water surface profile under free conditions. The velocity of the water in an underground stratum is the product of the permeability of the stratum times the hydraulic gradient. Most hydraulic gradients are quite small. Many have a slope of less than 0.1%.

Most of the time the shallow groundwater follows the ground contours. The deeper groundwaters, especially those in rock and under clay beds, move in response to the hydraulic gradients in their respective formations,

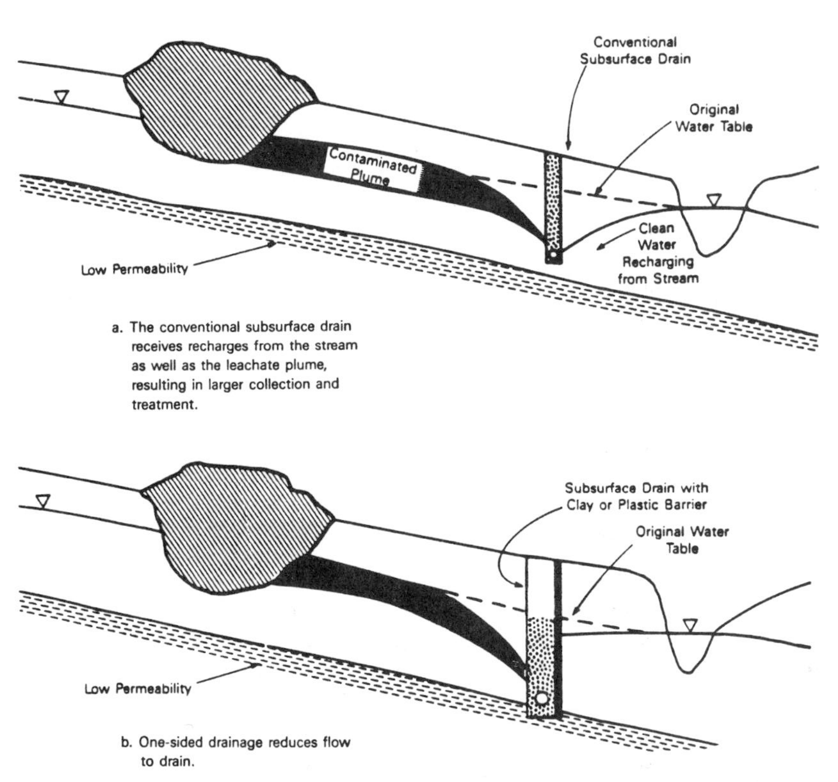

Figure 3.2 One-sided drain. (a) The conventional subsurface drain receives recharges from the stream as well as the leachate plume, resulting in larger collection and treatment; (b) one-sided drainage reduces flow to drain. Source: *U.S.EPA Site Remediation Manual.*

and they often move in a direction different from the shallow aquifer. The direction and rate of movement of groundwater in fractured rock are difficult to predict or determine. The movement through fractures may not follow the form of the equation given previously, and it is also extremely difficult to determine the amount, size, and hydraulic interconnection of the rock fractures.

While trenches and drains are generally thought of as being relatively shallow, at depths under 20 feet, a new drilling technique using a specially designed flexible well-drilling shaft shows promise as a new technology for site remediation. This new technique enables the driller to control the drift of the shaft, up to 90° from the original direction. When this "well" is installed, it can be gravel-packed and used as a deep collection device, or, if conditions permit, it can be used to aerate the water in the formation, providing oxygen for aerobic or methane for anaerobic biological growth.

The number of construction materials required for a drain or a trench are few. Perforated pipe, a few different grades of gravel, and sand are generally all that is required for construction of the drain or the trench. Geotechnical membranes, specially designed fabrics, are also used to help keep fine soils in place, helping to keep the drain or trench open. Well-constructed drains and trenches are built in vertical layers around the perforated pipe with the gravel and coarse material installed immediately adjacent to the pipe and the finer graded materials toward the outside. This type of construction is known as a filter. The purpose of a constructed filter is to prevent the fines in the soil from washing into the pipe and creating a void, or in silting in the pipe.

If barrier membranes are used in the construction of a trench, special considerations may be necessary. Water will seep into a drain under a constructed barrier, and any apparent head differentials observed between the "open" and the "closed" sides of a drainage trench as shown in the previous figure represent the extra hydraulic friction in the flow path the water has to take to get into the drain.

3.5 WELLS AND REMEDIATION

There are as many different types of opinions about the best combination of well, screen, and casing to use on a particular site as there are well drillers and geologists. At least half of them are right at any one time. The selection of the right sized well and material for casing and screen is dependent upon the water chemistry, the purpose of the well, the soil formations, the contaminants present, and the chemical interactions between the casings and the contaminants.

Wells have several principal uses in remediation systems: the extraction of fluid for recovery, the extraction of fluid for monitoring, and the reinjection of fluid into the formations. The fluid may be water or air. The engine for accomplishing the extraction or injection may be a centrifugal pump, a diaphragm pump, an ejector, or even an air lift.

Wells are installed in sizes from 2″ for monitoring wells up to 18″ and larger for recovery and production wells. On a remediation site, the most common sizes will be 4″ and 6″ wells. The larger diameter wells are installed when the formation has a high permeability and when the driller seeks to limit the velocity of the water around the well and the packing. While the 2″ well is generally a monitoring well, a recently developed small diameter pump is now allowing these wells to be used for some remediation activities. The larger sizes are most common for recovery and production wells at service stations.

3.5.1 DRILLING MUDS AND WELL DEVELOPMENT

Many drillers use drilling muds to keep the hole open during drilling and to help bring drill cuttings to the top of the well head for removal. There are a number of types of drilling muds, from plain water to organic muds. There are an equally large number of opinions regarding the use of these muds for any type of recovery effort. Many geologists do not see the need for such muds, and correctly argue that anything introduced into the drill hole will interfere with the quality of the sample collected from the hole. The argument is strong and persuasive. If not removed from the hole during the well development, drilling muds can cause plugging of the formation and low well yields. However, there are other geologists who believe that the proper use of drilling muds when coupled with good technique will insure adequate removal of the muds from the borehole, with no loss in performance of the well. The jury is still out on this question.

In highly contaminated formations where there is the possibility of many interferences from the drilling muds, many geologists prefer to use potable water as the drilling fluid, or if it is available, they will use an air/rotary drilling rig. The air/rotary rig has the disadvantage of being poorly suited for use in non-cohesive soils where the well could tend to cave in. In non-cohesive soils, the casing can only be advanced if muds are used to keep the casing open.

The well must be cleaned and developed when it is installed. The cleaning process can range from continuous pumping of the well until it runs clear (free of fines) or it can include such procedures as jetting (running a hydraulic jet down the inside of the screen to remove the drilling muds and to reseat the packing around the well). Wells drilled into clay formations are difficult to develop because the volume of water available in the well is very low. In that instance, the driller should be instructed to bail the well as often as possible over the course of several days to attempt to get most of the muds and fines from the well.

3.5.2 WELL TYPES

For the purposes of monitoring and extraction there are three major classes of wells; the differentiation between these classifications is characterized by their construction technique. These well types are differentiated in Figure 3.3. Type I wells are routinely used for routine groundwater monitoring where the possibility of surficial contamination is not a concern. Type I wells are backfilled with the soil that came from the well installation. Any contamination in the soil will find its way down to the well.

Type II wells are the most common for use at retail outlets and other locations that could introduce surface contamination to the well. Type II wells have a grout seal above the screened interval, and a bentonite clay seal between the grout seal and the well screen. Because of the cost of installing a bentonite clay seal and a grout seal, Type II wells are more expensive than Type I wells.

Type III wells are the most costly to construct and are generally only used in those situations where one has to drill through surface contamination to investigate the formation underneath. The Type III well is really two wells. A shallow well of large diameter is drilled to the bottom of the formation. The well is cased and bottom-grouted. During installation of the bottom grouting, the casing is lifted to form a mushroom around the bottom of the casing to insure adequate sealing. Once the first well is completed, a second well is drilled through the bottom plug to insure that the potential for contamination is minimized. A bentonite and a grout seal are also used to insure that the contaminants do not reach the lower formations.

3.5.3 WELL SCREENS AND WELL PACKINGS

Two very important components in the construction of a well are the well packing and the well screen. The function of the well packing is to help filter out soil from the formation and prevent it from plugging the well screen. The packing is also to help insure the smooth flow of water to the well and to prevent the soil around the well from becoming dislodged and entering the well. Depending upon the type of well and the purpose of installation, the well packing is generally a coarse sand or fine gravel. This material generally has a permeability greater than that of the well formation. The geologist installing the well selects the packing by examining the soil from the formation. The packing is installed around the well screen.

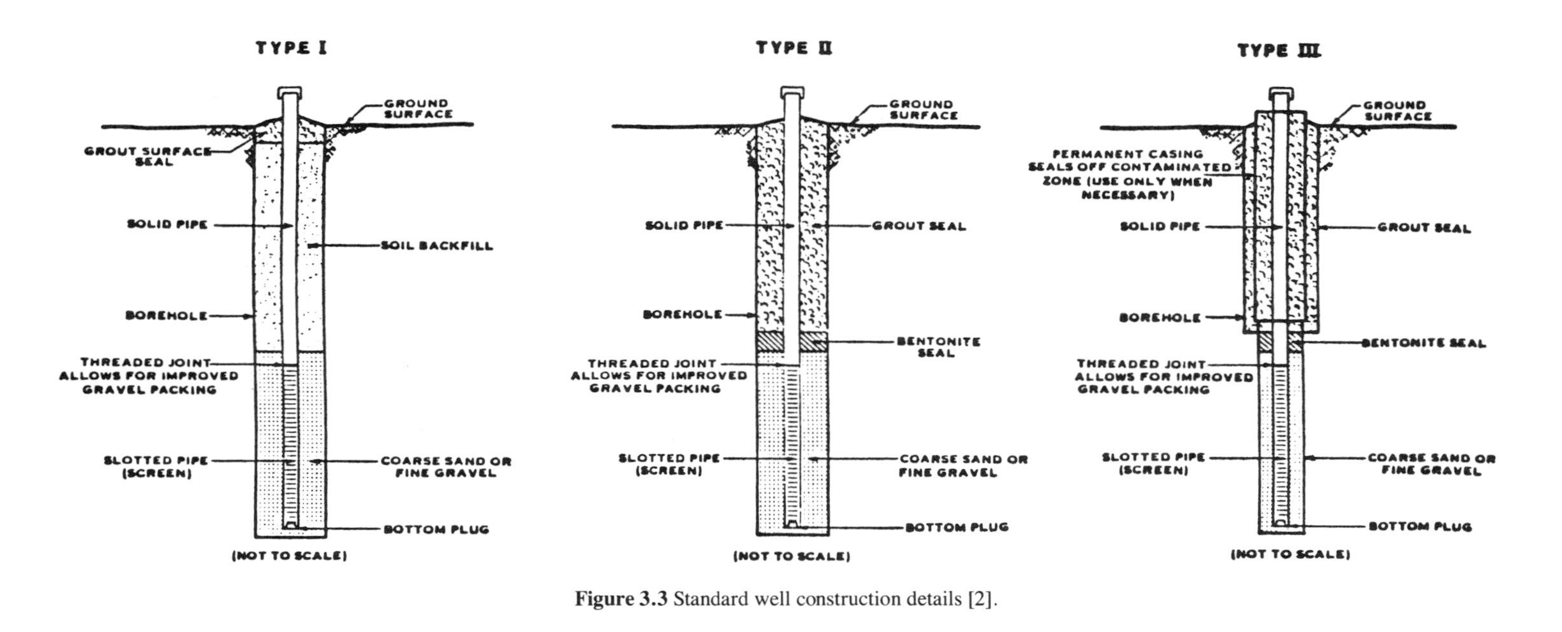

Figure 3.3 Standard well construction details [2].

A well screen is a length of pipe with fine slots cut into it or a frame around which evenly spaced wires are wrapped. A recovery well may have one or more screens. The number of slots and the thickness of the slots determines how fine a particle of soil will enter the well, and determines the effective open area of the screen. It is impractical for a well driller to construct a well packing large enough to filter out all the fines that could enter the well. Screen open areas range from 20% to 90%. Slot sizes range from about 0.01 inches upward. The screen open area and slot size is selected by the driller at the time of well installation. The selection of the right well screen is important to the adequate performance and long-term operation of the well.

The materials of construction of the well screen are important too. Some drillers and state agencies think that the only acceptable material for well screen construction is stainless steel. Polyvinyl chloride, Teflon, and polyethylene are also available and are well-suited to well applications.

Polyvinyl chloride or PVC is perhaps the most common material used for well construction today. It is resistant to many chemicals, is tough, and machines well, and it will not react with metals. For many wells it is the ideal choice for the well pipe, casing, and screen. PVC does have the disadvantage of being susceptible to chemical attack from high concentrations of petroleum products such as gasoline and diesel fuel. This attack causes swelling and softening of the pipe. Wells with 0.01″ diameter slots have been reported to swell shut in gasoline. To overcome this problem, slot sizes 0.02 inches and larger have been recommended.

PVC pipe is generally put together with methylene chloride and vinyl chloride monomer as a solvent/glue. These compounds show up in water analyses, and their presence can be of concern to environmental regulatory agencies because the compounds are alleged carcinogens. (Some companies and state agencies are insistent that glue joints not be used in putting together PVC pipe.) For the purpose of most remedial work, PVC pipe is a preferred alternative, and the moderate effects of chemical attack from gasolines are acceptable.

Teflon and polyethylene (PE) pipes are more chemically resistant than PVC, but are much harder to machine. Under the right circumstances, both will act as an organic sponge, and where analyses for trace organic chemicals are being performed, they should not be used. Polyethylene and Teflon well screens are available, but both are more expensive than PVC.

Stainless steel is chemically resistant to organic compounds, but it has the distinct disadvantage of being very expensive and hard to machine. Consequently, stainless steel pipe and well screens are very expensive when compared to PVC. Unless there is a specific need for stainless steel well screens, its installation represents an unwarranted expenditure for most circumstances.

3.5.4 HYDROGEOLOGIC MODELING

All remediation efforts using wells, trenches, or drains should be accompanied by some hydrogeologic modeling. The amount and sophistication of the modeling effort (and the expense) will depend upon the complexities of the specific site, and the abilities of the modeler. At a minimum, the original and modified surface contours should be run to determine the effects of the installation and the pumping of the wells, or the changes that will come about from installation of drains.

The use of a groundwater model is essential in sizing the pumps for a particular well. Even a very simple model, such as that shown in Figure 3.4, will help one visualize the movement of groundwater. Figure 3.5 shows another view of the same hydrogeologic flow pattern. Both these plots were made on inexpensive, commercially available software, and both were taken from a cleanup on a service station.

In both views, an unusual formation exists in the groundwater. The unexplained valley in the groundwater contours suggests a strong movement due to a highly permeable natural drain. Further investigation showed the cause of the valley to be related to a local storm sewer at the site. In this instance, the remedial alternative was pumping and treating the groundwater, first to remove free product, then to recover water from the plume. The final solution to control the groundwater involved several pumps, and the final surface water elevations during the pumping looked like those shown in Figure 3.6.

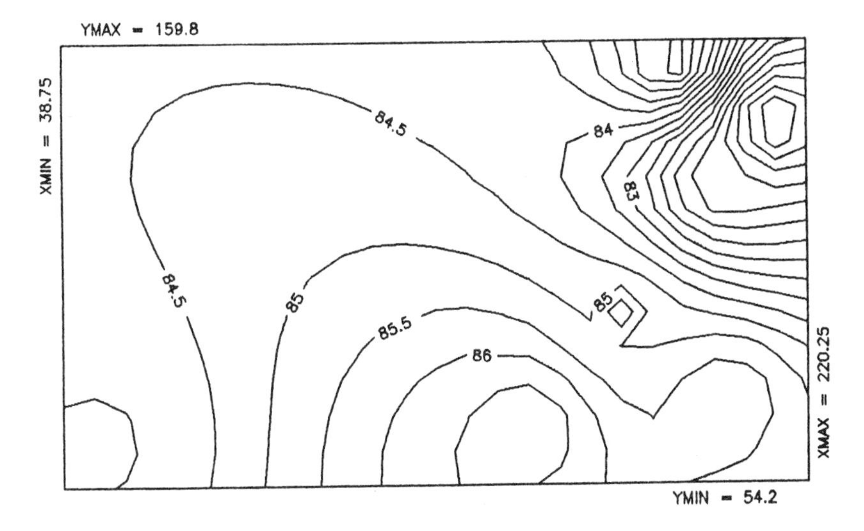

Figure 3.4 Groundwater profiles at a service station.

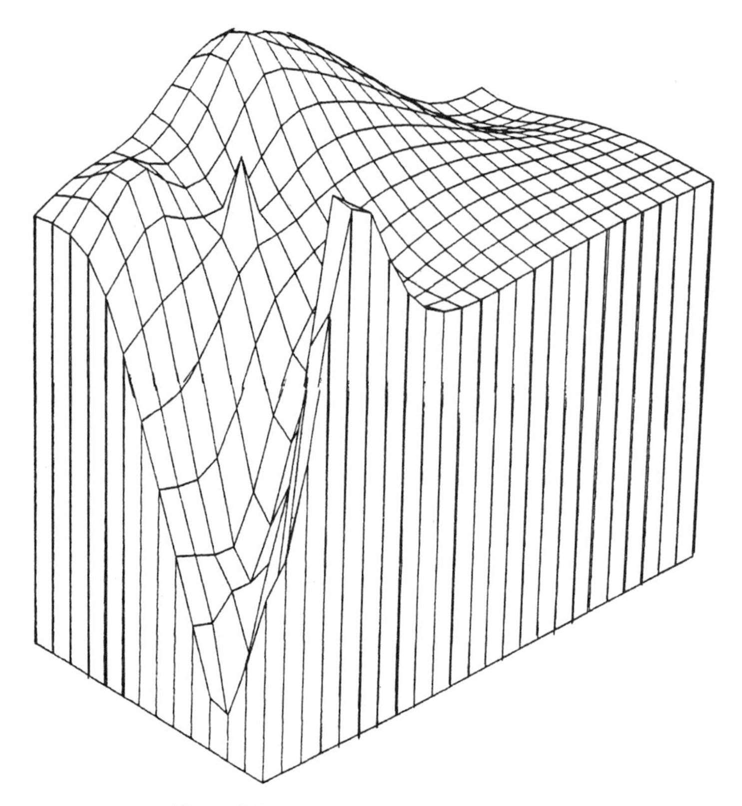

Figure 3.5 Another view of the same profiles.

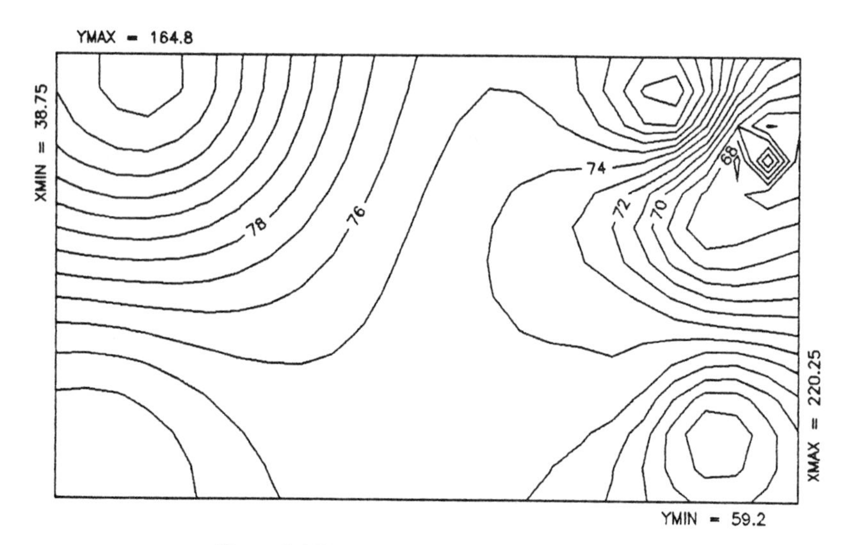

Figure 3.6 Water surface contours after pumping.

In Figure 3.6 it should be observed that the valley shown in Figures 3.4 and 3.5 has been removed, and the water elevations had become influenced by the cone of depression created by the pumping of wells in the area. In this instance, the cone of depression was designed to be limited to the property line because of local contamination from other service stations across the street. The water (and the free product) now flow into the cone of depression where it will be collected by the recovery pumping system.

3.6 PUMPING SYSTEMS

There are three types of pumps in common use for collecting groundwater samples: centrifugal pumps, positive displacement pumps, and eductors.

A centrifugal pump employs one or more curved vane impellers that rotate on a center shaft. The pumps have a curved head-discharge relationship as shown in Figure 3.7. The pump performance is specific to impeller diameter, impeller shape, and speed of rotation.

The motor can be on top of the well or can be submersible. Many recovery wells use submersible motors. The smallest motors currently available are just under 2″ in diameter and fit down a monitoring well. Electric motors have minimum cooling requirements, and submersible pumps should not be installed in situations where there is a low flow and high discharge head.

Centrifugal pumps can also be used on top of a well to draw water up to the surface. The maximum theoretical lift of any vacuum is 33 feet of water at sea level. In practice, a theoretical lift much above 20-27 feet should not be used. The application of a partial vacuum to lift water up to the surface will remove some of the volatile compounds in the well and change the water chemistry.

Positive displacement pumps have head-discharge relationships as shown in Figure 3.8. The most direct type of pump uses a diaphragm or a plunger to physically displace the water from the well. Principal operating variables include number of strokes per minute, stroke length and bore (displacement) of the pump, and the construction of the pump. Some positive displacement pumps use air as the prime mover in the pump, and for these pumps, the air pressure and quantity to the diaphragm is also a consideration in sizing the pump. For mechanical pumps, the theoretical limit on pump discharge may be dependent upon the strength of the materials of construction.

The element that makes most positive displacement and diaphragm pumps work is the check valve. When the pump cylinder or diaphragm is ascending, the pump is drawing water through a check valve on the suction side of the pump. When the cylinder or diaphragm descends, the inlet check valve seats tight and the discharge check valve opens, allowing water to leave the cylinder. If the check valve is not compatible with the fluids

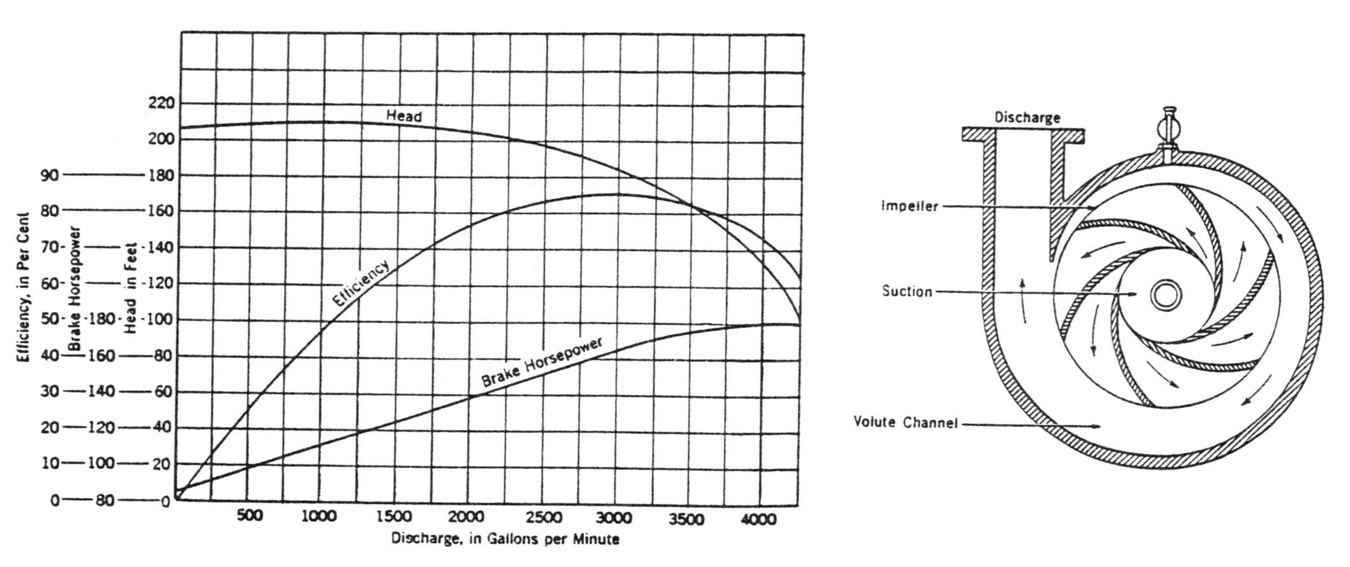

Figure 3.7 Centrifugal pump construction and curves. Source: *Water and Wastewater Engineering* [3].

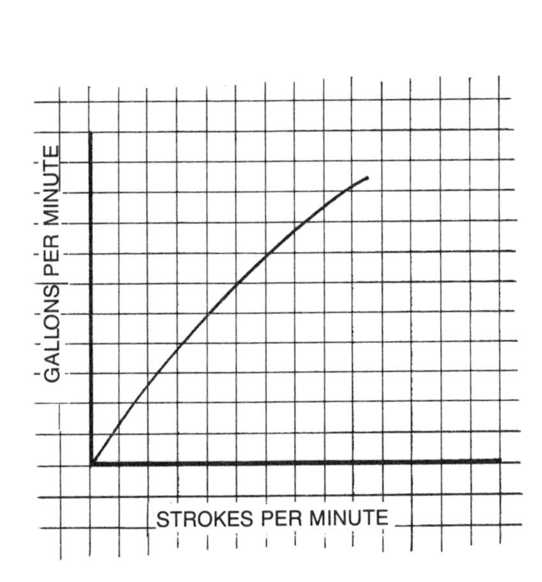

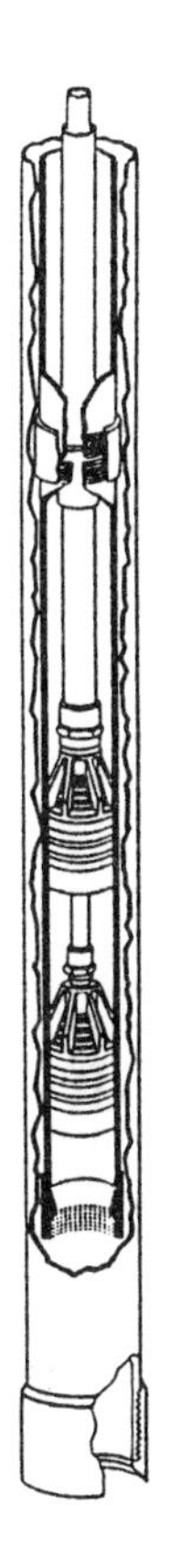

Figure 3.8 Positive displacement pump and curves.

pumped, or if it fails to seat properly, the pump performance may decrease substantially.

Figure 3.9 illustrates the eductor. The eductor uses a cylinder to collect the water. A series of check valves admit the water but close tightly when the cylinder is pressurized. The water is then forced up through the eductor tube into the piping and to the top of the well. Since air is generally the pressurized fluid, the sample can be lifted to almost any height depending upon the air pressure supplied.

3.6.1 TOTAL FLUID PUMPING VS. RECOVERY PUMPING

Most petroleum-contaminated sites have free product as well as a dissolved plume. The free product is often referred to as a non-aqueous phase liquid (NAPL). The free product is generally recovered separately because

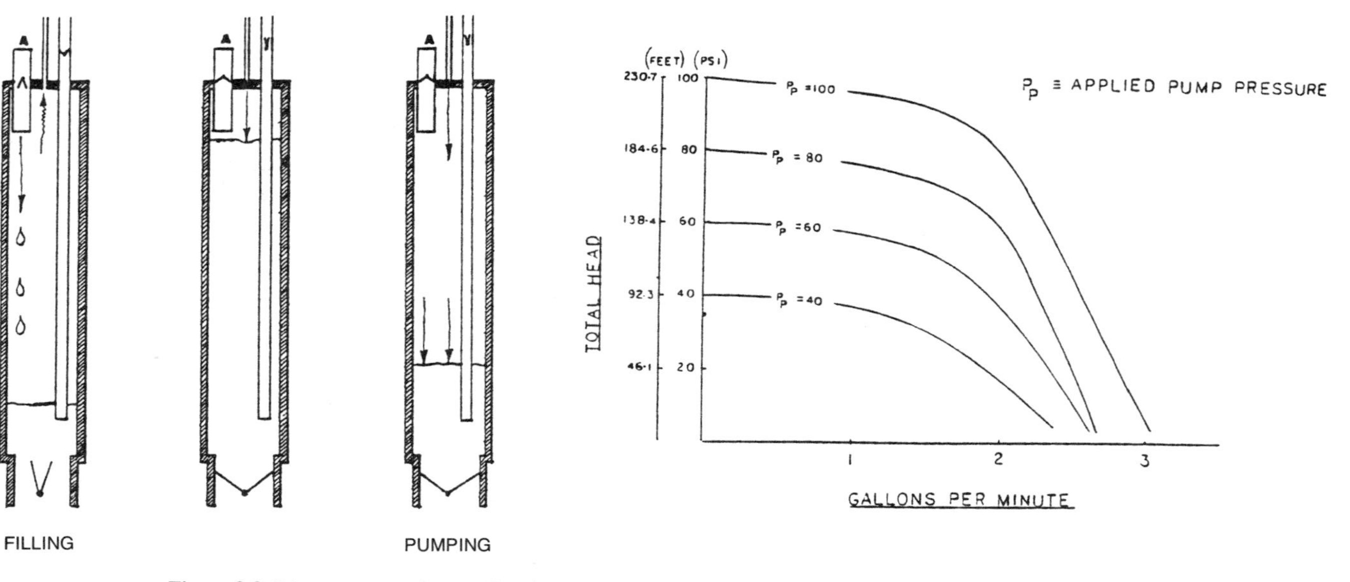

Figure 3.9 Eductor system. Source: Westinghouse Environmental Systems and Eductor Systems.

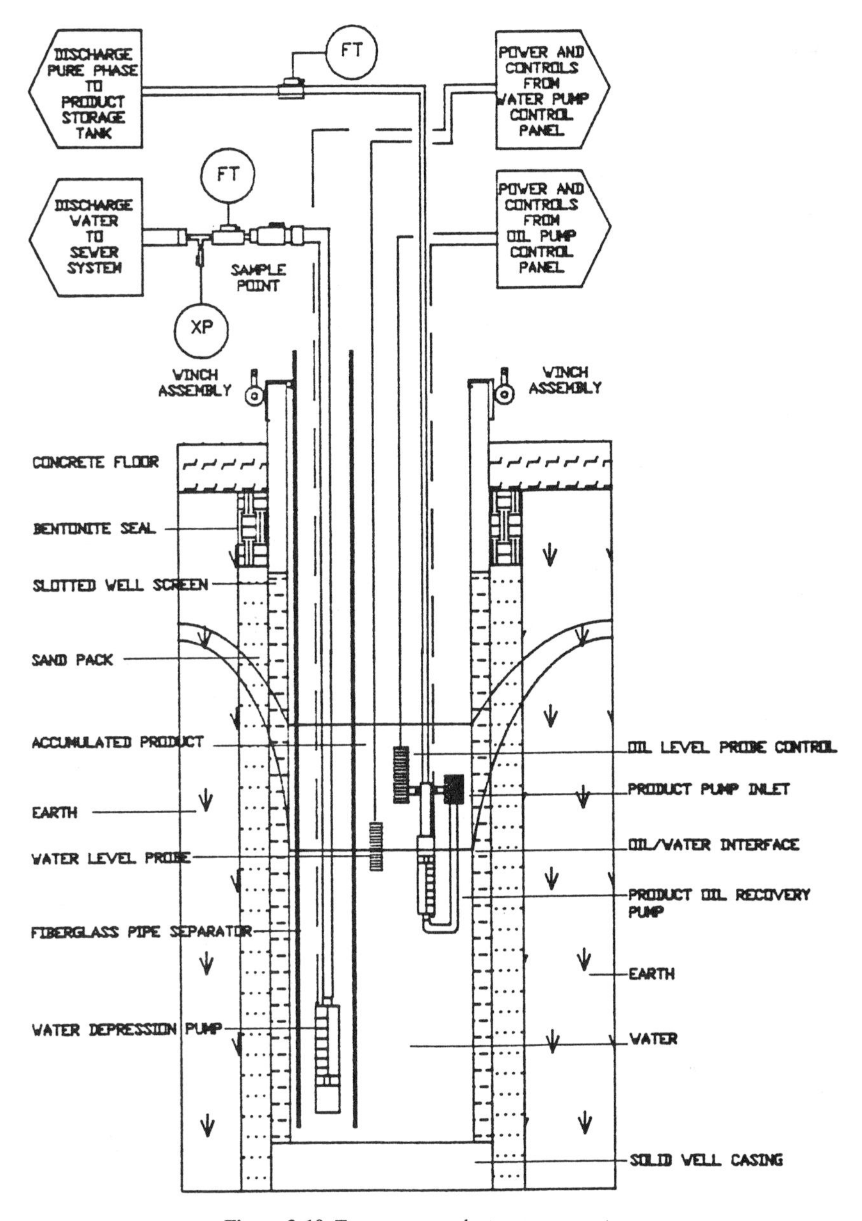

Figure 3.10 Two pump product recovery system.

it is easier to treat separately. If the free layer product is thick enough, it is a relatively simple matter to install a pump directly into the free product and pump it to a recovery tank. If the free product layer is a few inches or less, direct recovery becomes more difficult, and a second pump is used to depress the water level and increase product thickness. This is shown in Figure 3.10.

The two pump system makes recovery easier, but it has the disadvantage of having a lot of controls and pumps down in the well. The installation of a two pump system in a 4″ well can be accomplished only with some difficulty, and in smaller wells, not at all. One of the practical problems with two pump systems is the fluctuations in water levels in the well. The two principal sources of fluctuation arise from pump discharge not exactly matching the well yield for each fluid, and from seasonal water table fluctuation. Summertime drops in an area's water table can leave pumps high and dry.

Total fluid pumping uses one pump that pumps all the fluids in the well. The equipment cost of the one pump system is less than half of the cost of the two pump system. Total fluid pumping will sometimes increase the amount of dissolved gasoline and organic materials in the fluid pumped, and sometimes this renders the water more difficult to treat. In these situations, the pumping may create an emulsion. Since the emulsion often does not easily re-separate, it will carry through to the water treatment equipment where it may be more expensive to remove. The use of a good de-emulsifier such as a chemical additive, or a physical de-emulsifier such as a packed bed coalescer or a fibrous-type coalescer, can separate the water and the fuel, reducing the load on the treatment equipment at extremely modest capital cost.

Most often, a site that has a gasoline contamination problem does not have a dense non-aqueous phase liquid (DNAPL) plume. However, there are those sites where a combination of fuel oils or other materials may have leaked out of a tank and formed a DNAPL layer on the bottom of the aquifer. These plumes are among the most difficult to remove because they are often relatively thin, are more viscous than the water, they move more slowly, and because of their density they are frequently found in an area of the aquifer where it is physically difficult to remove them. Recent investigations have indicated that the DNAPL materials may, in some instances, be recovered by a two pump system where a water movement pump is located above the DNAPL recovery pump.

3.7 WATER TREATMENT SYSTEMS

One of the most important considerations in water treatment is the water chemistry. While this section is not meant to be a comprehensive review of the aspects of water chemistry, it will mention some of the principal problems that are frequently encountered in site remediation and then will look at some of the principal ways in which petroleum products are often removed from the water.

3.7.1 WATER CHEMISTRY

The presence of certain inorganic compounds in the water can make the removal of inorganic chemical much more complex. One of the chief compounds causing trouble in strippers and aerators is iron. In its dissolved form, in the +2 valence state, it is quite soluble. When iron is brought into contact with air, as in a stripper, it oxidizes to the +3 valence state, and forms Fe_2O_3, or rust. The rust is a natural precipitate of the water, and will deposit on everything it touches. Most people are familiar with the iron staining that occurs in a toilet bowl or a sink; when water containing iron is run through a packed tower or stripper, the deposits can plug the stripper.

Other inorganic cations in water are also important. Dissolved calcium, magnesium, or manganese can react with nutrient phosphates added during a remediation effort to form a plug that may seal a well or blind off an entire formation. These substances are "dissolved rock," and they may, if conditions are right, form extensive deposits on and inside treatment equipment.

Carbon dioxide is soluble in water. When carbon dioxide dissolves, it forms a weak acid and the bicarbonate ion. The presence of carbonates indicates a potential for chemical deposition in and on the treatment equipment. Bicarbonate ions react with calcium, magnesium, and manganese (and iron) in rock formations to solubilize them. If the pH of the water is changed, or if the water is heated, the ions will precipitate as bicarbonates or as carbonates, forming insoluble deposits that can plug pipes and other equipment.

Hydrogen sulfide is a gaseous contaminant of water that has the smell of rotten eggs. It is often present when the water is at or near zero dissolved oxygen content. When water containing H_2S is aerated, the H_2S is released, and the surrounding community then notices the smell. It is more of a nuisance that needs to be handled than a problem that needs specific remediation. In high concentration, however, H_2S is lethal, as concentrations over 5 ppm by volume will temporarily paralyze the olfactory senses, rendering them useless as an early warning detector of higher concentrations. At remediation sites, the H_2S can be of greater danger to a driller or mechanic entering a manhole to service a pump than it can be to the community at large.

The pH of the water may be a problem in certain communities. If the natural pH of the water is below 6.0 standard units (SU), the water may be considered corrosive and will attack metal tanks, and metal pump parts. If the pH is above 8, the water is slightly alkaline and will indicate a slight tendency to deposit dissolved minerals on heating elements, pipes, and in pumps. If the pH is above 10, the water contains alkaline material that will quickly plug pipes and treatment equipment.

High levels of alkalinity and acidity in water are an indirect indicator of the dissolved minerals in the water and a direct indication of the degree of difficulty one may have in changing the pH of the water if it is required for other types of water treatment.

3.7.2 PHYSICAL SEPARATIONS

Two types of gravity separation methods are used to separate gasoline from water, or diesel fuel from water. Both are based on the difference in specific gravity between the fluids. The most common type of gravity separation is by decanting or quiescent settling. The fluid is placed in a tank and the separation takes place naturally. The tank does not need to be a "dead" tank but can have flow through it if that flow is small enough not to cause eddy currents and turbulence.

If the diesel or gasoline is emulsified in water, the mixture will look milky and will separate extremely slowly, if at all. If the mixture smells like gasoline or diesel but is clear without a separate layer, the petroleum is dissolved and will not separate after long periods of standing. Additional treatment is required before the petroleum can be removed.

One of the most popular types of flow through separators is the API separator, a large baffled tank with a large surface area. The tank may have an open or closed top. The baffles at the top of the tank prevent short circuiting of the surface flow, and help collect the product. When the level of floating product is sufficiently deep, the floating product is drained off for recovery or disposal.

A second type of separator is a coalescer. A typical coalescer design features a series of closely spaced baffled plates, generally set at 45° to the flow and 90° to each other. The slight turbulence the fluid encounters by changing direction is generally sufficient to force many of the suspended and emulsified particles to bump into each other and form larger droplets that will be large enough to float. This type of coalescer is generally followed by a smaller surface separator. The coalescer generally can achieve free oil concentrations in its effluent of less than 20 mg/l.

Serfilco, in Chicago, Illinois, manufactures a cartridge-style filter/coalescer that is excellent. This coalescer physically resembles an industrial thread cone and will remove emulsified product to levels less than 10 mg/l. The principal drawback to the unit is that it also acts as a filter, and may blind when turbidity in the water is high. As a protection, a bank of inexpensive, disposable, cartridge-style filters is generally added ahead of the coalescer. The unit can be used to achieve physical separations where the difference in specific gravity is as low as 0.05 units.

3.7.3 FILTRATION

Filtration is used for separating solids from liquids. The solids most commonly encountered may be sand or clay fines or chemical precipitates from reactions. Filters used at a typical site might be of the cartridge style, or might be self-backwashing sand filters. For example, a clayey soil frequently contains a number of fines that must be removed prior to aeration so that they will not plug the aerator. The filter used for this type of separation should have a filtration rate of approximately 2 gallons per minute per square foot of filter area. The filter should be a sand filter of the self-backwashing pressure type.

Filters can be relatively high-maintenance items unless they are properly instrumented. Even then, they will still require some operation and maintenance. Cartridge filters must be changed when the pressure drop across the filter exceeds a predetermined value. It is substantially cheaper to change a filter than to attempt to clean out a plugged pipe that may result if a filter is not in the system.

3.7.4 STRIPPERS AND AERATORS

One of the methods of removing dissolved gasoline from the groundwater is to use a stripper or an aerator. A stripper is generally a tall column packed with Rasch rings (which resemble a doughnut), shaped wires, or shapes that have a high surface to volume ratio. Figure 3.11 shows a typical arrangement for a stripper and packing materials. The water containing the hydrocarbons to be removed (stripped out) is introduced at the top of the tower, and air is introduced near the bottom for countercurrent flow. The volume of the air is about 100 times greater than that of the water. Depending upon the removal requirements, the height of the tower is between 15 and 30 feet. The air leaving the tower is saturated with water vapor and contains modest amounts of hydrocarbons, and droplets of water. The water leaving the bottom of the tower may contain less than 1 microgram of petroleum per liter (>1 ppb) which will meet most regulatory criteria for discharge.

Aerators are another form of stripper. The aerator is an air/liquid transfer device into which air is blown or stirred. The blowing aerator can be compared to a child blowing bubbles in a glass of water. The stirring device can be compared to a blender. In the first instance, the liquid containing the hydrocarbons is pumped continuously through a shallow (depth $<$ 15 feet) tank into which air is bubbled. The tank may have a mixer

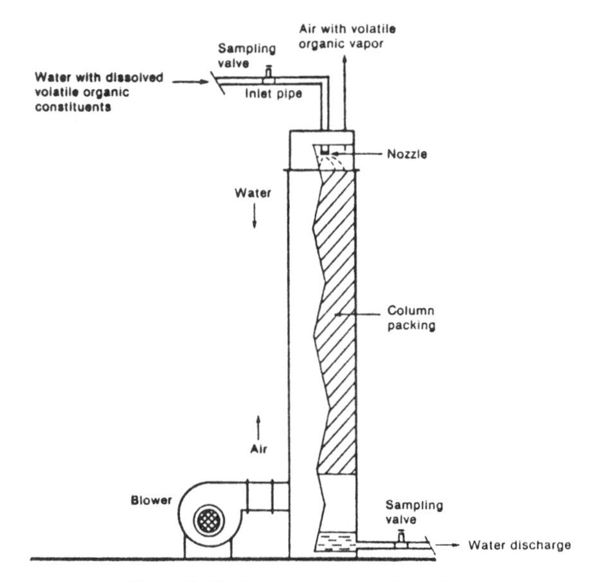

Figure 3.11 Stripping tower and packing.

or may rely on the mixing motion of the air. This configuration is identical to that found in a number of municipal wastewater treatment systems. In the second process, air is blown and beaten into the liquid by a rotating propeller mixer and air sparger located under the mixer propeller.

Either type of aeration system can be an effective alternative to a stripping tower; neither type is adversely affected by iron or suspended solids in the water. A recent report in *Ground Water Monitoring Review* [4] reported that a $200,000 pretreatment system was installed just to remove the solids and iron to protect a $20,000 stripping tower. The cost of an aerator that could have performed the same job as the stripping tower and which would have been unaffected by the solids was on a par with the cost of the stripper. There are several different methods of treating water to remove volatile compounds and to achieve a high-quality effluent at a modest price.

The discharge from the stripping tower may contain droplets of water. These will often fall to the ground around the site, creating a nuisance by spotting cars and equipment. They should be removed by a demister. The demister is a low pressure drop device that filters out or coalesces out the water droplets in the exhaust from the stripper. Aerators do not need demisters.

Many states, especially New Jersey, North Carolina, and South Carolina, have adopted strict emission controls regarding the discharge of hazardous air pollutants and volatile organic chemicals (VOC's). The chemicals found in gasoline include benzene, ethyl benzene, toluene, and xylene, which are considered hazardous air pollutants by some states. All states recognize the potential for a stripper to discharge VOC's. Benzene is a hazardous air pollutant that the states most often seek to regulate. It might not be unusual for a state to require a secondary treatment device on the back of a stripper to eliminate the discharge of benzene. The regulations vary from state to state.

Georgia, for example, has two regulatory criteria on the emissions of volatile organic compounds (gasoline). If the remediation site is within a non-attainment area for volatile organic compounds (VOC's), the limit will be a maximum of 15 pounds per day. Outside the metropolitan areas, the limit is 550 pounds per day. North Carolina has a 100 pound per day limit on the amount of hazardous air pollutants that can be emitted, and generally calculates gasoline emissions as containing 25% hazardous compounds.

Other types of emission control may also be of concern. Recently, a remediation design was developed for a service station in the Atlanta area that considered odor control. The site was located next to a restaurant that could have been impacted by the chemical odors coming off the stripping tower.

In order to control the atmospheric emissions, additional equipment is required. Appropriate control devices include fume or vapor incinerators, carbon columns, and vapor condensers. Each one has advantages and disadvantages, as will be discussed later.

3.7.5 CARBON ADSORPTION

Carbon adsorbers have two potential uses in emission control from contaminated service station sites. The first use is as a direct method of water or wastewater treatment; the second is as a means to control vapor emissions. In this section the use of carbon in water treatment will be discussed.

Activated carbon is an effective means of removing small quantities of hydrocarbons from water. The principal mechanism is by adsorption. Activated carbon has an extremely porous structure, similar to that of a microscopic sponge. When organic compounds come in contact with carbon for a long enough time, the organic materials are adsorbed into the holes in the "sponge." When the "sponge" is full, no more adsorption can take place. To be truly effective, the carbon must have a large surface area with many adsorption sites; a carbon bed with a small particle size will have a much greater surface area than a carbon bed that has a large particle size.

Activated carbon is manufactured from wood, coconut shells, and other organic materials that have been heat-treated in a closed atmosphere without oxygen. Each carbon has slightly different absorptive properties for a given chemical. While generalizations can be made regarding the specific capacity of a carbon from any manufacturer, the best information about the capacity of a specific carbon to adsorb organics from a specific waste stream is gained by running a pilot adsorption isotherm test on the material.

A typical adsorption isotherm is shown in Figure 3.12 [5]. The amount of carbon required for a specific removal can be calculated from the slope of the adsorption isotherm and the experimental constant. This is expressed as

$$x/m = K * C^{1/n}$$

where

x = mass of hydrocarbon adsorbed
m = mass of carbon used
K = experimental constant
C = equilibrium concentration
n = slope of adsorption isotherm line

The computations are chemical-specific and are additive. The final carbon dose for removing gasoline from the water would be calculated by examining each of the isotherms for carbon and the specific chemicals and adding each of the chemical dosage requirements together to get a total.

Carbon is not a very efficient medium for adsorption of chemicals from water. Typically it requires 1 kilogram of carbon to adsorb 8 grams of petroleum hydrocarbons. The adsorption of high levels of hydrocarbons is not recommended because of the large amount of carbon required to attain the removal levels generally required.

When a carbon bed has adsorbed the maximum amount of material that it can hold, it is considered spent. Untreated water entering the bed will gradually be treated less efficiently and the effluent concentration will rise gradually and then sharply as the bed approaches exhaustion. For a typical carbon bed, exhaustion can happen fairly quickly, so when the chemical concentration in the effluent begins to rise, the carbon is considered exhausted and a new bed is brought into service. Figure 3.13 shows a typical carbon breakthrough curve.

Carbon can be regenerated by thermal roasting, but this is not practical in quantities of less than a ton per day. Sometimes, carbon can be regenerated by steam heating to remove the volatiles. This process is economic on a scale of several hundred pounds per week. Both processes will extend the life of the original carbon beds, but both will decrease the overall removal efficiency with each regeneration. The decision to use one type of regeneration or the other depends primarily upon the scale of the operation and the ability of the chemicals to be removed from the carbon. Laboratory tests are a must for this type of decision.

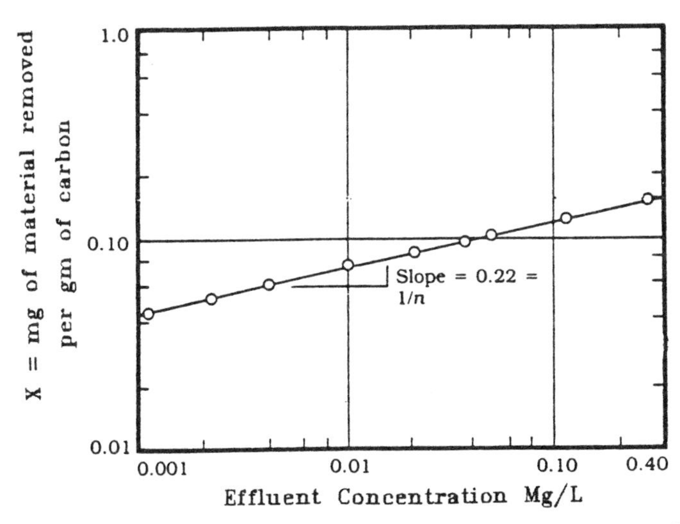

Figure 3.12 Typical adsorption isotherm [5]. Source: Reynolds, Tom D. 1982. *Unit Operations and Processes in Environmental Engineering*. Belmont, CA: PWS Publishers.

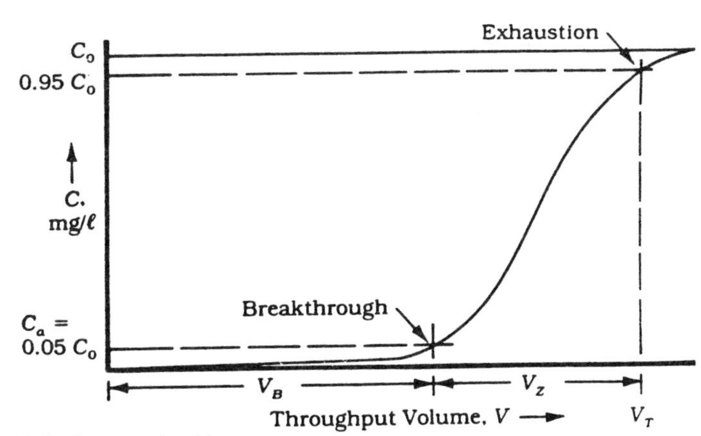

Co = Initial Concentration

Vt = Total Volume

Vb = Volume at Breakthrough

Figure 3.13 Breakthrough in a carbon bed [5].

If the carbon cannot be regenerated or the scale for regeneration is too small for on-site work, the carbon is collected and returned to the manufacturer for regeneration or containerized for disposal. The principal cost advantage of regeneration, where the manufacturer will accept the carbon, is cost. The cost of carbon replacement is not materially different than that of new carbon. The cost of disposal of the spent carbon can be as much as the cost of the new carbon, because the spent carbon is often regarded as a hazardous waste.

3.7.6 BIOLOGICAL TREATMENT

Aboveground biological treatment is only slightly different in principle than *in-situ* biological treatment and landfarming. Aboveground biological treatment is conducted in a tank as a process operation. The water is then returned to the ground or is discharged.

Biological treatment is a complex process that requires operation if it is to be conducted successfully. Unless other considerations dictate, biological treatment is not the process of choice for the treatment of gasoline-contaminated waters. The same comment does not apply, however, to the treatment of soils or *in-situ* bioremediation because the purposes of the treatment and the operating techniques are somewhat different.

Biological treatment is a process where microorganisms are induced to feed on complex substrates (gasoline and petroleum compounds), converting them by oxidation to carbon dioxide, water, nitrates, sulfates, and cell growth. Biological treatment does not take place with a single organism but requires a number of different types of organisms in a mixed culture. Nutrients must be added to the culture so that the approximate carbon: nitrogen: phosphorus ratio (C:N:P ratio) is between 100:5:1 and 100:20:5. The fundamental governing equations for biological treatment, which illustrate the character of the oxidation process, are:

$$C_xH_yO_z + O_2 \xrightarrow{\text{microorganisms}} CO_2 + H_2O + \text{energy}$$

$$C_xH_yO_z + H_2O + NH_3 \xrightarrow{\text{microorganisms}} \text{cellular material}$$

$$+ CO_2 + H_2O + \ldots$$

For each pound of organic material destroyed, approximately 0.77 pound of new cells is produced. These cells must be periodically removed (wasted) from the system. Depending upon the amount of growth taking place, the rate of sludge wasting may be several pounds per day or several pounds per week. The waste cellular material is generated in a liquid stage, generally under 2% solids, which must be de-watered and tested prior to disposal.

Biological systems are temperature-sensitive. Even large bio-treatment systems are sensitive to temperature shock, and all systems will be less effective in waste removal with a rapid change in temperature. For every 10°C increase in temperature, the rate of organic reaction doubles. For every 10°C decrease in temperature, the rate of organic reaction halves. A doubling in removal rate does not bring about a doubling of removal efficiency; with a rapid change in temperature over a 24-hour period, the sudden rise or fall of the liquid temperature can bring about a marked decrease in treatment performance.

In cold weather, biological treatment system removal rates sometimes drop to 10% of the summertime values as the water approaches freezing. In order to operate a small biological treatment facility effectively in cold weather, it should be enclosed in a heated building, and the water should be kept as close as possible to a constant temperature.

Biological treatment systems are also sensitive to the concentration and type of materials in the influent stream. A sudden increase in organic loading can upset a bio-treatment system, causing the microorganisms to go into shock and lose treatment efficiency. Certain organic chemicals found in gasolines are resistant to bio-degradation. The effective treatment of all chemicals in diesel fuels and gasolines may make a treatment system prohibitively large.

Gasolines are deficient in phosphorus and nitrogen, and these materials must be added as nutrients in approximate proportions to the concentration of organic materials in the system. The nutrients used in a typical system are ammonia or sodium nitrate and phosphoric acid or a sodium salt of phosphoric acid.

Biological treatment systems are also pH sensitive. Since the systems produce CO_2 and organic acids, there is a mild tendency for the pH to decrease, but the buffering capacity of the water is generally sufficient to overcome this tendency. If the pH gets below 6, however, a biological treatment system may be severely inhibited.

Biological treatment systems often produce a highly nitrified effluent as they convert ammonia NH_3 to nitrate NO_3. Nitrates in the groundwater are a cause for regulatory concern, as the nitrate concentration limit in public drinking water supplies is 10 mg/l. Phosphate in the effluent can react with calcium in the water to form calcium phosphate, a crystalline solid material that has the ability to form deposits capable of plugging groundwater formations.

In summary, for groundwater treatment, biological treatment above ground is not a recommended technology, as it is expensive and somewhat tricky to operate. Other cheaper alternatives are available.

3.8 INCINERATION

As a remedial alternative, incineration is quite acceptable in that the end product is a clean soil free from organic material and suitable for replacement in the same hole. Florida and South Carolina and many other states have developed extensive programs for incineration of gasoline-contaminated soils. South Carolina actively encourages incineration. Other states such as Tennessee recommend that the soil, once removed from the contamination site, be treated in an asphalt plant. A number of remediation contractors have developed mobile incineration systems. This disposal technique is gaining much wider acceptance. Portable incinerators can be transported by truck and can be set up on-site in a few days. The principal problems associated with using a transportable incinerator are scheduling the incineration, shutting down all other operations at the site during remediation operations, transporting the incinerator, cost, material handling, and obtaining the permits to operate the incinerator.

In order for a soil to be treated by incineration, it must first be excavated, stockpiled, and prepared for incineration. This generally requires a sheltered area with a roof and floor or a membrane cover beneath and on the soil stockpile. Side walls may also be necessary if the soil is fine or has a tendency to dust during preparation. Where volatile emissions are a concern, states have required the stockpile of soil be covered with an inflatable membrane cover, and all excavation operations must take place beneath this "tent." The air in the "tent" must also be treated to reduce volatile organics.

When the soil is taken from the stockpile, it is blended to achieve uniform consistency and BTU content and then fed into a vibrating mixer/screen where rocks above 2″ diameter are screened out. These materials are generally not processed in the incinerator but are sent directly to a landfill or are otherwise decontaminated.

After the initial screening, the soil is placed on a belt feeder and may be dried before incineration. The incineration takes place in two steps, a volatilization step and a destruction step. The volatilization of the soil takes place in a rotating kiln at temperatures up to about 700°F. The gases from the volatilization are then processed in a secondary combustion chamber at significantly higher temperatures ranging between 1600°F to 2400°F. The secondary combustion insures the degradation of all the organics in the soil. The remainder of the process train is comprised of air pollution control equipment that may or may not include a baghouse and a scrubber for any acid gas that may be generated.

After the soil exits the primary combustion chamber, it has been decontaminated and is cooled and returned to the excavation. Figure 3.14 shows a large incineration system with a capacity of 16 to 20 tons per hour [6]. The figure is presented for information purposes only, and any equipment used on a service station site would be much smaller [7].

Other elements that may be involved in the incineration of soil include the treatment of scrubber water and the handling of the ash. Scrubber water is usually a mild sodium hydroxide (NaOH) solution adjusted to a slightly alkaline pH so that it does not remove the carbon dioxide from the waste gas, but only the acid gas (HCl) generated from the combustion of chlorine, which may be in the soil. The scrubber water is used to control dusts and cool the ash pile. The disposal of the scrubber water may be of concern, and in some instances, states have required the scrubber water to be run through a carbon adsorber to insure that there are no residual organics. Before it is returned to the earth, the ash is tested for compliance with regulatory criteria. If the extractable metals in the soil are not excessively high, the excavation is backfilled using the decontaminated soils. If, however, the soils fail the regulatory criteria, then they must be re-incinerated or other treatments must be applied.

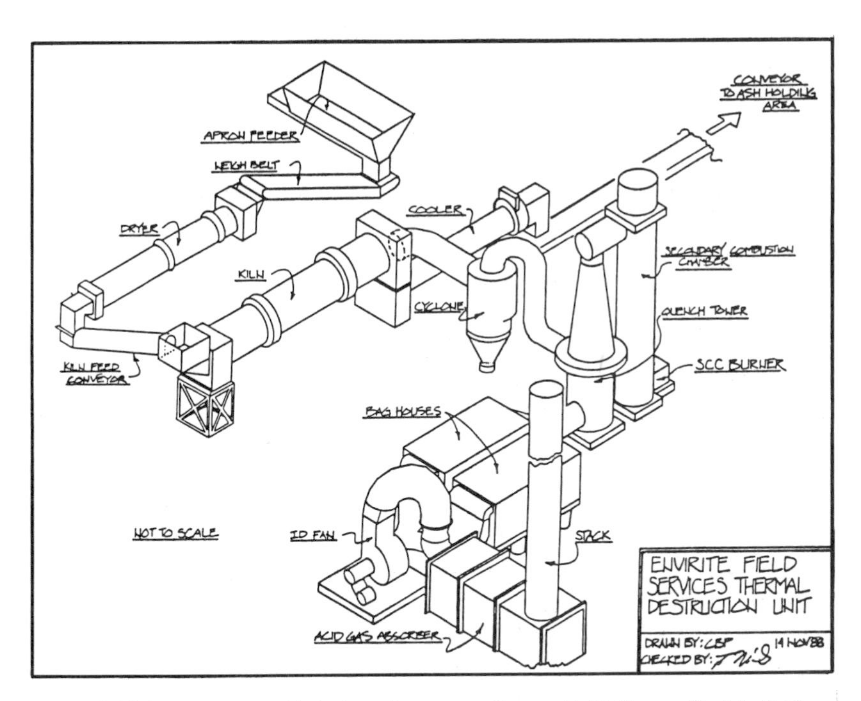

Figure 3.14 Large transportable incineration system. Source: T. McGowan, Envirite field services.

One of the primary concerns of the regulatory community is the prospect of the release and oxidation of heavy metals found naturally within the soil. Chromium found in the soil may be converted to a higher valence state by incineration/oxidation. In this higher valence state, the toxicity of chromium is substantially enhanced as is the compound's solubility and mobility. If after incineration the soil is considered a hazardous waste, there will be little choice except to process the soil to stabilize it and reduce the potential for chemical migration. If the soil is declared a hazardous waste, even a hazardous waste landfill cannot accept it without first treating it to reduce the leaching potential of the chromium. This would probably double the total cost of the treatment.

3.9 SOIL VENTING

Soil venting is gaining popularity because it offers high removal rates for volatile components in the gasoline contamination in the soil and it is relatively inexpensive to construct and operate. A basic soil-venting installation consists of several screened wells connected together with piping and a vacuum pump. Depending upon the needs of the local environmental authorities, the air stream from the back of the vacuum pump may require treatment to control hydrocarbon emissions from the well. Soil venting is particularly effective in the rapid cleanup of contaminated sites within limits that will be discussed below.

Soil venting utilizes the low viscosity of air and the high volatility of gasoline to remove contaminants from the soil. The gasoline has a relatively high vapor pressure and diffuses into the soil vapor until it reaches saturation at the equilibrium. At saturation, the air in the soil contains approximately 25,000 ppm of gasoline. When the concentration of gasoline vapor in the soil is lowered, gasoline diffuses into the soil air from the liquid.

By moving air through the soil, the soil vapor is extracted and the concentration of the vapor in the soil is reduced. The gasoline "evaporates" or diffuses into the air. The extraction of air from a contaminated soil area has the ability to remove several hundred pounds of gasoline per day from a site; even under optimal conditions, the recovery of free product in a liquid pumping system would remove less than 10% of the material removed by a vapor recovery system. This is illustrated by Figure 3.15 [8] and Figure 3.16 [9].

Vapor extraction systems will not efficiently remove compounds with low vapor pressures. Compounds that have a low volatility will remain in the soil. Any product recovered through vapor extraction will have a disproportionately higher distribution of volatile materials when it is compared to the composition of the gasoline spilled. The material remaining in the soil will have a higher concentration of the lower-volatility compounds, which

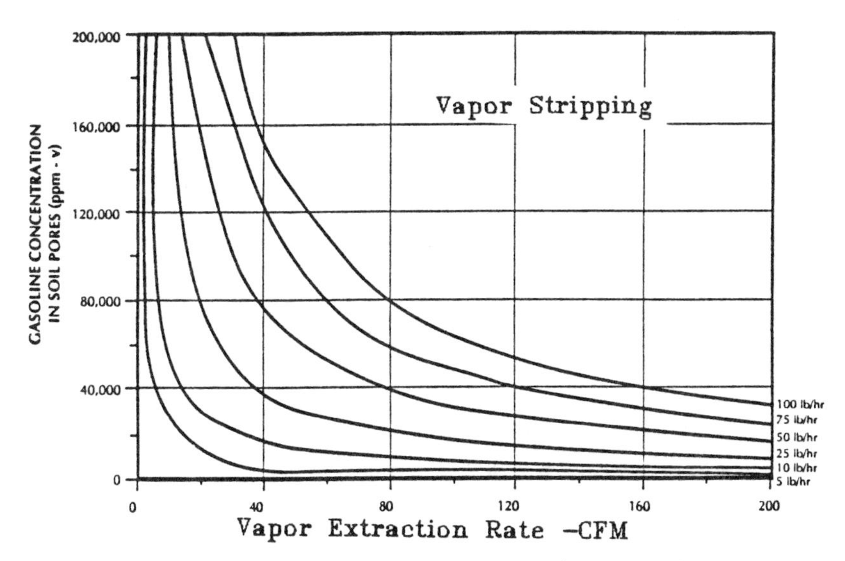

Figure 3.15 Soil vapor extraction rate. Source: U.S.EPA seminar on groundwater treatment.

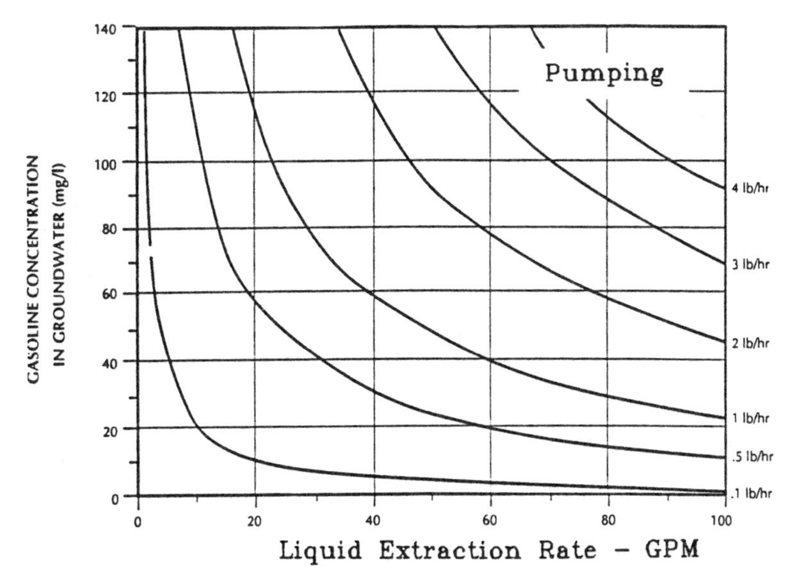

Figure 3.16 Liquid extraction from soil. Source: U.S.EPA seminar on groundwater treatment.

are hard to remove and hard to biodegrade. In a typical recovered gasoline, the compounds isobutene, *n*-butane, isopentane, *n*-pentane, hexane, and ethylbenzene will dominate; similarly, the compounds in the soil will contain higher concentrations of 1-hexene, C-12 aliphatic compounds, 1,3,5-trimethylbenzene, methylcyclohexane, xylene, and other compounds.

The effective radius of influence of a vapor extraction well is limited, and the vapor moves through the well in proportion to the porosity of the soil, its moisture content, permeability, and particle size. Figure 3.17 shows the typical vapor extraction profiles from a recovery well in a sandy New Jersey soil [10]. The effectiveness of the vapor recovery well and removal system will depend upon the relative permeability of the soil, the soil moisture content, the grain size of the soil, the soil porosity, and the vacuum applied to the soil. A recovery well may have a radius of influence for effective removal of soil vapors of about 10-20 meters if soil conditions are right and much less if the soil consists of impermeable clays or silts.

The sealing of the surface over a recovery site by a concrete slab will cause the air resistance in local wells to be increased because the air withdrawn from the well cannot be recharged from the area under the concrete slab. It must be withdrawn from a greater distance, increasing the length of the path across which the air must move, and increasing the pressure drop required to move the air into the well.

Vapor extraction wells have been designed using a vacuum of up to 22″ of mercury, but the more common figure is a maximum of around 6″ of mercury. Air flow rates in a vapor extraction well range from approximately 15 cfm to as high as 100+ cfm. The most common figure used in design is about 25 cfm. The vapor extraction process causes the water level in the well to rise (mound) in response to the vapor withdrawal.

Vapor extraction will dehydrate the soil. The water in the soil wets the soil particles and displaces the hydrocarbons because the soil has a greater affinity for the water than it does for many of the hydrocarbons. When the water is removed, the hydrocarbon adheres to the soil, entering some of the soil pores and physically binding itself to the surface of the particles. Consequently, the hydrocarbon is not as available for evaporation, and the diffusion rate of the hydrocarbon into the soil air is reduced. When the soil air concentration is reduced, the hydrocarbon removal rate drops off.

In order to optimize the removals of a soil extraction system, the system must be allowed to rest periodically to restore the soil moisture and the hydrocarbon diffusion balance. After a relatively short resting period, the hydrocarbon removal resumes at a higher level than when the extraction was stopped. Figure 3.18 illustrates the effect of extraction and the effect of halting the extraction process temporarily and allowing a recovery period for the soil. Note that in the broken curve in Figure 3.17 the soil extraction resumes at a greater rate than before it was discontinued, and the initial and

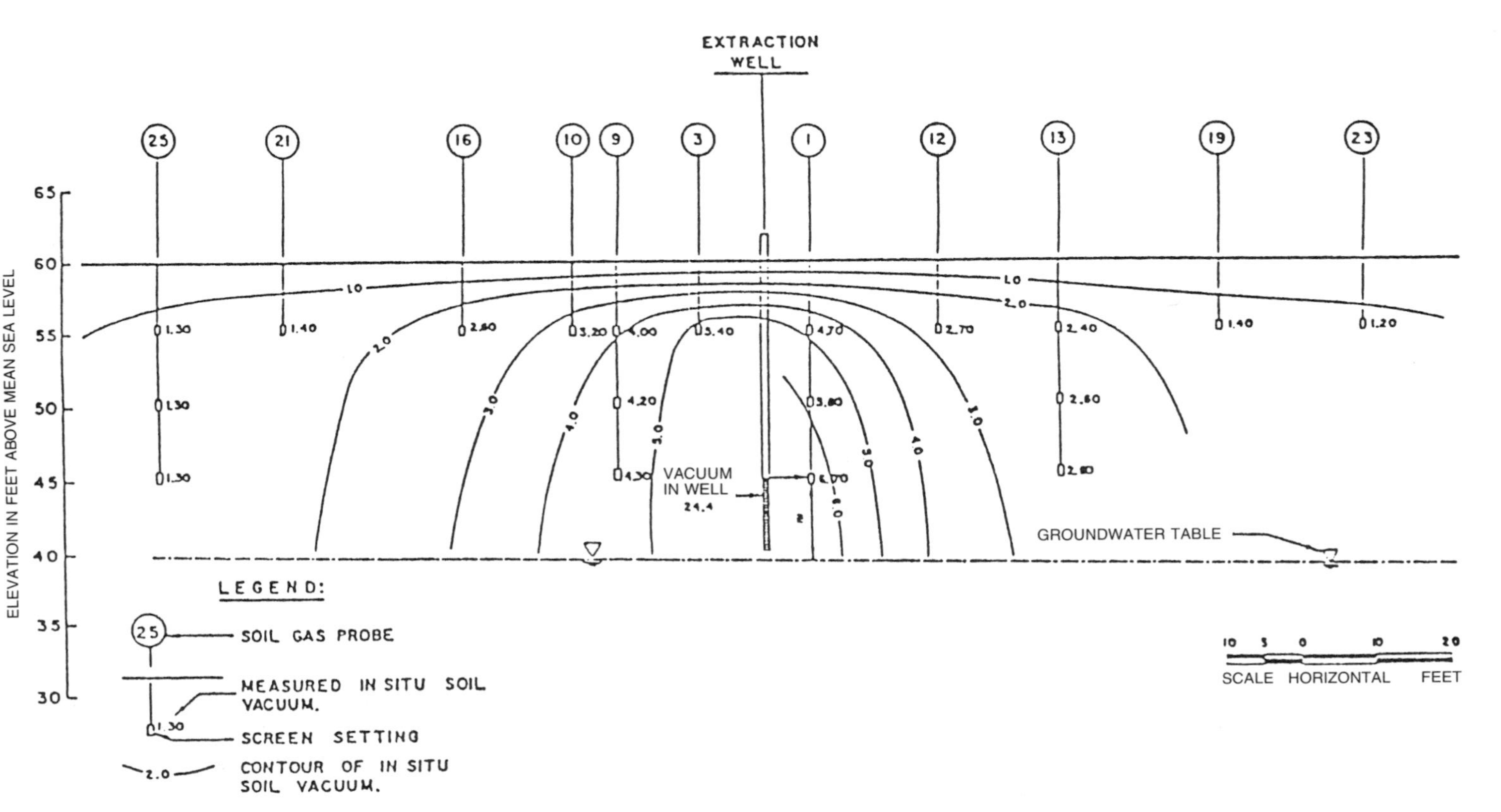

Figure 3.17 Cross-sectional *in-situ* soil vacuum contour map (inches of water).

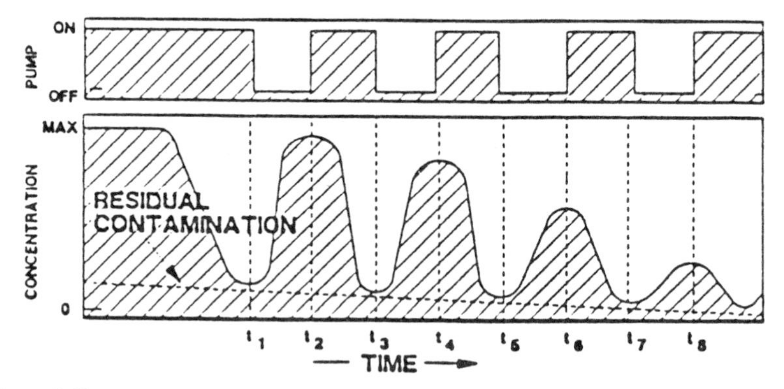

Figure 3.18 Decrease in soil hydrocarbon levels by periodic system resting. Source: U.S.EPA seminar on groundwater treatment.

resumed extraction rates both increase as the flow in the well field is re-established. This suggests that the optimal method of operating an extraction system is through periodic pumping and rest cycles [11].

3.9.1 VAPOR EXTRACTION SYSTEMS ANCILLARY EQUIPMENT

Special caution must be exercised in the development of a vapor extraction system because the vapor extracted often exceeds the lower explosive limit for gasoline vapors. The equipment utilized should be explosion-proof. Any motors should be totally enclosed and unventilated, and all blowers should be made of non-sparking metals such as aluminum, brass, or bronze. If a vapor incinerator is required to reduce the emissions, a water seal and a flame arrestor are strongly recommended. The equipment should be installed outdoors, inside a fenced area with warning signs and no smoking signs on the installation fence. If the air in the blower is within the explosive range, the temperature rise in the blower may also be a consideration, as excessive temperature rise in the event of a restriction in the air flow may cause accidental ignition. High exhaust temperature alarms and a shutoff switch on the blower may be warranted.

In the event that the vapor product cannot be discharged into the air, it must be adsorbed, condensed, or incinerated. Carbon is much more efficient in adsorption in the vapor phase than in the liquid phase. Figure 3.19 shows a typical adsorption isotherm of benzene. As before, the adsorption isotherms are chemical compound specific and additive. The isotherm data from individual compounds must be examined and the results added together. If the vapor being stripped is saturated with water vapor, it will inhibit the efficient adsorption of the carbon. In order to insure that the air is unsaturated, heat must be added or additional air must be introduced. In many cases, the air exiting the vacuum pump has been warmed sufficiently to decrease the relative humidity in the air.

If the air is to be incinerated, several types of commercial incinerators are available. Some are even available for rent. Catalytic combustors are generally good for hydrocarbon removals up to about 6000 ppmv and removal efficiencies up to about 90%-95%. The catalytic combustor operates at an ignition temperature of 500°F to 750°F and uses a platinum- or nickel-impregnated catalyst to burn the fuels in the air stream. Catalytic combustion is slightly more energy-efficient than regular combustion because the ignition temperatures are almost 1000°F lower. However, with a heat regenerator on the incinerator, much of the claimed economic advantage disappears.

If the concentration of the volatile vapors in the exhaust from the extraction system is much beyond 5500 ppmv, the volatiles concentration may be too high for catalytic incineration. The heat available from the combustion of the fuel will exceed the ability of the catalyst to dissipate heat without warping, melting, or losing its properties. Many contractors prefer to dilute the vapor by adding outside air to the influent vapor stream. If the gasoline vapor levels are above the lower explosive limit, and if dilution air is added, a fire or explosion may result. If the vapor levels cannot be safely and economically lowered, incineration may be required.

Incineration of the vapors will reduce the volatile emissions in excess of 98%. The incinerator is relatively forgiving of fluctuations in concentra-

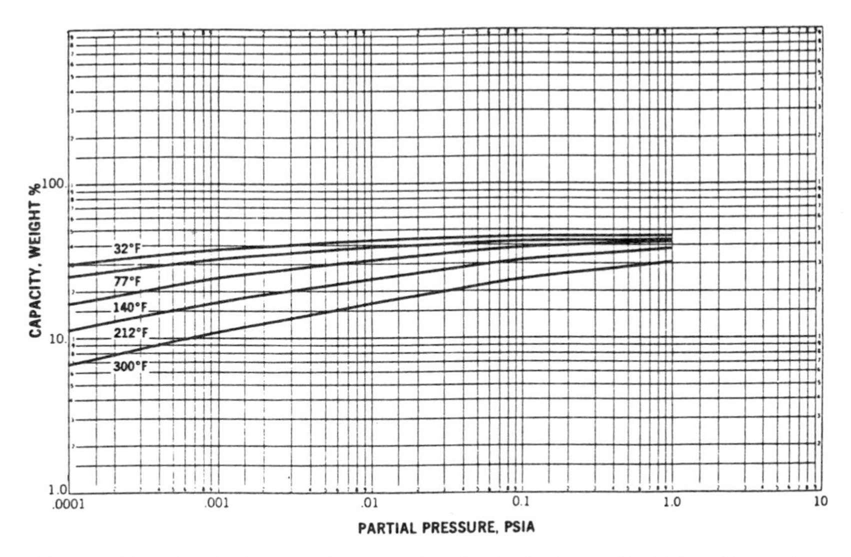

Figure 3.19 Carbon isotherms for benzene adsorption in air. Source: The Calgon Corporation.

tion, and depending upon the control mechanisms, it can go from concentrations of a few hundred ppm to 25,000 ppm in a matter of seconds. The incinerators and the catalytic combustors use a continuous flare or pilot to keep the temperature up and insure adequate combustion. The most efficient fuel for the vapor combustor or the incinerator is natural gas. Propane and fuel oils can be used at a greater expense.

Vapor-condensing systems have been used for a number of years for reducing the hydrocarbon emissions from a loading rack on fuel terminals. The equipment is made by a number of manufacturers. The equipment is essentially a large capacity chiller that takes the air temperature down below the condensation temperature of the volatiles in gasoline, below about 10°F. The condensation systems work very well but may have a problem with freezing, as they are designed to work primarily on systems without much water vapor in them.

The vapor condensor emits a cold plume that is less buoyant than the surrounding air and can create higher ground-level odors than a warm plume. Since dispersion is the major consideration in evaluating potential air complaints and compliance requirements, the cold plume problem may require some special attention. Cold plume problems can be overcome by adding heat to the plume or by building a taller stack. Because the vapor-condensing system is a large compressor/refrigeration system, it is between three and five times more expensive than a comparably sized incinerator.

3.10 BIOREMEDIATION

This section will discuss two similar types of bioremediation alternatives currently in use. The first one is landfarming, the aboveground system in which the contaminated soil is seeded with manure and fertilizers and is allowed to biodegrade the petroleum compounds in it. The second system is *in-situ* biodegradation, where nutrients are added to the soils and contaminants in place to achieve biological degradation of the pollution.

3.10.1 LANDFARMING

When underground storage tanks are removed and contamination is discovered, the soil is excavated and stockpiled for removal or for landfarming. The process encourages aerobic degradation of the contamination within the soil. For a typical contaminated soil pile, the treatment procedure will run approximately like the following:

(1) The soil is first spread in uniform lifts generally less than 1 foot thick over an impermeable membrane to control run-off and re-contamination of the ground. (Parking lots serve quite well for this purpose.)

(2) If required, cow or horse manure and fertilizer are added to the soil to improve the nutrient balance and to increase the available microorganism mass in the soil. These materials are then disked into the soil. (The fertilizer used for the nutrient addition has an N:P:K ratio on the order of 10:2:2.) The principal nutrients are adjusted to achieve a C:N:P ratio of approximately 100:5:1.

(3) When the manure and the fertilizer are added, the soil is worked and tilled until all the material is combined. This generally requires several passes with a tractor dragging a disking machine. The soil is then watered and left alone.

(4) The soil is sampled at periodic intervals and the hydrocarbon content, the soil moisture, soil nutrients, and temperature are evaluated and plotted.

(5) The fastest degradation takes place in the summer when the temperatures are the highest. In winter the biological degradation virtually ceases. The fastest biodegradation will occur when the soil is maintained with a constant moisture, neither too wet nor too dry. The exact moisture content of the soil should be determined in a laboratory or by a soil scientist.

(6) For optimal removal rates to be maintained, the soil should receive periodic moisture and should be tilled at intervals between six weeks and three months.

Within a period of a few months, all the hydrocarbons have been consumed and the soil is ready for reburial or other uses. Some of the best regulatory guidance available on landfarming was developed by the State of North Carolina Underground Groundwater Protection Division[1].

When the concentration of volatiles in the soil falls below regulatory limits, the state may permit it to be buried or disposed of in a non-hazardous waste landfill.

Enhanced landfarming with the addition of manure can increase the quantity of microorganisms in the soil mass, and this form of biodegradation can accomplish in weeks what may require months or years without the nutrients and the manure[2].

The enhanced biodegradation process is quite effective in treating diesel fuels as well as gasolines. The higher levels of nutrients and microor-

[1]The document is available for $2.00 from North Carolina Department of Environment, Health, and Natural Resources (NCDEHNR), 441 N. Harrington St., Raleigh, NC, 27603, Attention: Paula Hall. Ask for the *Soil Remediation Guidelines*.

[2]Much of the work on surface biodegradation has been accomplished by the Environmental and Agricultural Schools at North Carolina State University.

ganisms have been found to degrade even highly resistant compounds such as pentachlorophenol and acenaphthalene and pyrenes.

3.10.2 *IN-SITU* BIOREMEDIATION

In-situ bioremediation holds the most promise for inexpensive remediation of petroleum-contaminated sites where contaminants are deep and other remediation methods are impractical. The field is still relatively new. As recently as ten years ago, soil textbooks stated flatly that microbes could not and did not exist in the deep soils beyond the vadose zone. The prevalent theory then considered any soil samples taken from a depth of beyond 20 feet as sterile. Today, we realize that organisms can and do exist in the deep soils, and that given the right conditions, they can be encouraged to develop the enzymes necessary to treat the contaminants that man has put there. As an example of this point, a researcher at a biological treatment conference was heard to complain about a lack of any toxaphine (a highly effective and toxic pesticide) with which to conduct his government-funded treatment and research experiments. The last plant that made toxaphine in the U.S. was shut down about ten years ago, and even when he sampled the soil at the plant he could not find the material in sufficient concentration or quantity to permit sampling and the performance of treatability studies.

In-situ bioremediation is an ideal process to follow vapor stripping of the soil. It promises the most effective treatment of soil contamination. Currently there are two approaches to bioremediation: aerobic treatment and anaerobic treatment. Anaerobic treatment, treatment conducted in a reducing environment, is currently being researched for the degradation of polynuclear aromatic chlorinated and non-chlorinated compounds; it is substantially more difficult to manage, and it has slower reaction kinetics than aerobic treatment. However, many compounds that are toxic or highly resistant to degradation by aerobic organisms can be degraded by anaerobic treatment. For the remainder of this chapter the discussion will be restricted to aerobic treatment.

3.10.3 PRELIMINARY CONSIDERATIONS

Microorganisms need a carbon source, nutrients, and oxygen to degrade hydrocarbons and change them into cellular growth, carbon dioxide, and energy. In the zone of contamination, the carbon source is in plentiful supply. Nutrients and oxygen are missing. The biodegradation will proceed naturally as fast as it can if one supplies sufficient quantities of nutrients and oxygen. The problem of supplying nutrients and oxygen where needed and in the quantities needed is the major challenge.

Before beginning a biodegradation process, one would do well to consider that the microorganisms will degrade anything that they consider food. The metabolism processes taking place are elementary, and the microorganisms will first consume the "foods" that are the simplest and require the least expenditure of energy on their part. If a particular strain of organism has to choose between a glucose molecule and a benzene molecule, the glucose will be degraded preferentially because it yields the most energy for the effort expended. A general rule of thumb in dealing with biological systems is, "Given any combination of time, temperature, and pressure, a microorganism will do exactly as it pleases." In that regard, it should be noted that the entire purpose of bioremediation is to provide the necessary conditions to get the microbe's purposes and ours to coincide – namely, to degrade the organics in the soils.

Water can contain dissolved oxygen, but only about 20 mg/l. In its simplest form, a molecule of carbon will require two molecules of O_2 to convert it to carbon dioxide. Similarly, four molecules of H_2 will combine with one molecule of O_2 to produce two molecules of water. Given a pound of a simple hydrocarbon found in gasoline, such as isopentane, which has the formula C_5H_{12} and which comprises up to 15% of some gasolines, the oxygen required to completely degrade the compound would be 4.879 pounds. At a saturated condition of 18 milligrams per liter, it would need 4.879 (pounds of oxygen required) * 453,600 (milligrams per pound)/18 (milligrams of oxygen per liter of water) = 122,950.800 liters of water, or approximately 32,500 gallons of oxygen-saturated water.

Even in a formation of moderate permeability, this requires a lot of water to move past a relatively small area. Groundwater flow is multi-directional, and it is almost impossible to direct a quantity of water to a specific location. Obviously, water saturated with air is not the optimal solution as a carrier of oxygen.

By adding hydrogen peroxide to water, the oxygen content can be increased. Hydrogen peroxide solutions up to about 500 milligrams per liter have been shown to be non-toxic to most microorganisms, and at that strength, the oxygen content is almost 26 times higher than that provided by air saturation, and the corresponding quantity required is slightly over 1250 gallons. While that is better, some professionals feel that it is still not enough. Hydrogen peroxide will react with dissolved iron in the water and oxidize it to produce ferric hydroxide, which is noted for its ability to plug wells and formations. Hydrogen peroxide is also relatively expensive to use, and in commercial concentrations (30%), it can cause severe burns to unprotected skin.

Drilling techniques exist to permit the installation of horizontal wells for direct aeration beneath contaminated soils. In the example above, if it requires 4.879 pounds of oxygen to biodegrade a pound of chemical, that weight of oxygen would be contained in 252 cubic feet of air. When transfer efficiencies are considered, it would require approximately 4800 cubic feet of air. While this volume may seem large, it is actually well within the range of an inexpensive commercial blower. In fact, the air can be delivered over a very short period of time, much faster than the microorganisms can assimilate the hydrocarbons.

3.10.4 AERATION SYSTEMS

There are two types of direct aeration systems: vertical wells and horizontal wells. The vertical well is a conventional well used for introduction of air into the formation. The horizontal well or drain can be drilled from a vertical well with special equipment that allows the shaft to be curved to the horizontal. It can also be bored from a caisson. The horizontal well is more expensive to construct than a simple vertical well, and the technology is available only from a limited number of drilling companies, but it is still substantially less expensive than the caisson.

In the caisson construction method, a shaft at least 6 feet in diameter must be sunk in the soil, and a horizontal boring machine lowered down to the desired level to advance the boring. Conducting any type of operations below ground in a shaft of moderate diameter is expensive. Specialized contractors, generally those with tunneling experience, are required for the work.

When air is applied to the vertical well, the effect is the same as blowing through a soda straw in a glass of water. If air is applied to the horizontal well, the effective aeration area associated with the well has a much more pronounced lateral spread. Figure 3.20 illustrates the concept of direct aeration of the soil [12].

With either aeration system, there is often a need to attempt to intercept the plume and treat it in place. The operation of a system of injection wells and recovery wells helps to contain the plume and make it move in the path of the aeration systems.

3.10.5 NUTRIENT ADDITIONS

Nutrients are required for bioremediation. Bioremediation proceeds slowly, if at all, without them. As noted above, the ratio of C:N:P will be somewhere around 100:5:1. The nutrients most commonly used are ammonia, an ammonium salt, or sodium nitrate and sodium phosphate. The use of sodium nitrate is discouraged because of the problem of introducing a potential contaminant into groundwater. Nitrates have been known to cause the "blue baby" syndrome in small infants, and the maximum

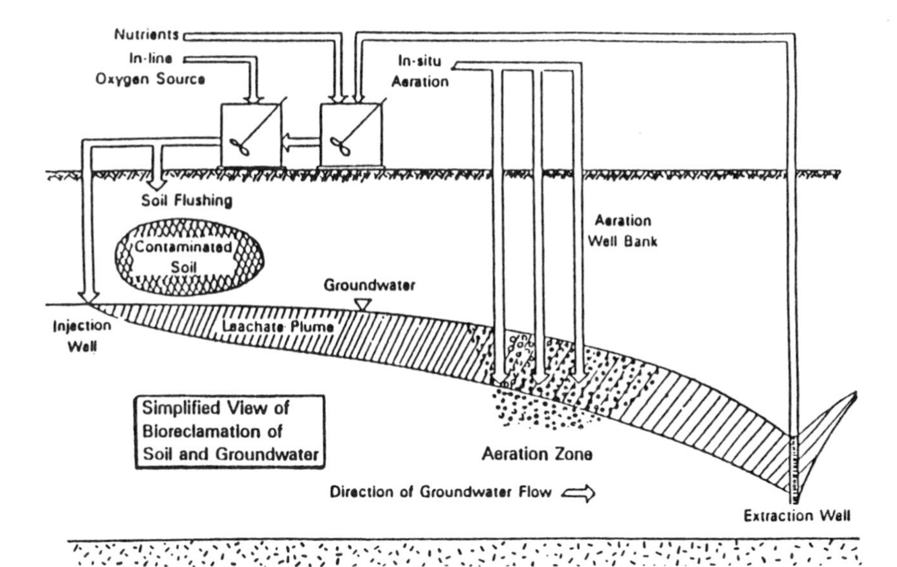

Figure 3.20 Direct aeration and treatment of a contaminated soil. Source: U.S.EPA seminar on groundwater treatment.

concentration of nitrate permitted in a drinking water supply is 10 mg/l. The use of a soluble phosphate can cause problems in a formation with limestone or calcium rock geology. Calcium phosphate is a highly insoluble precipitate, and its use has been known to reduce the hydraulic capacity of a formation by over 100 times in a matter of weeks.

Other problems that have been associated with the use of nutrients in an injection well include the plugging of the injection well. While many authors speculate that an abundance of nutrients and oxygen around the injection site causes direct plugging of the well by excessive bacterial growth, there is some evidence to indicate that the plugging may be primarily due to microbial generation of gas, and the trapping of that gas in the micropores in the soil surrounding the well [13].

3.10.6 SITE RESTORATION AND MONITORING OF BIOREMEDIATION

Some of the practical difficulties in developing a bioremediation system for a particular site are involved in dealing with contaminants in the deep vadose zone above the seasonal high water table. If nutrients, including phosphates, are to be delivered to the zone of contamination, the zone must be carefully defined and then the area water table must be raised to promote the degradation. This does not say that no biodegradation can take place without raising the local water table. It can and does, but it does so at a relatively slow rate because the nitrogen and phosphate nutrients needed for rapid degradation are contained in the delivery system (the water).

The nutrients must also be delivered to the zones in which the contamination exists. In the common case of a multi-layered formation, where the permeability is different in each layer, the rate of water movement through a sand layer may be several orders of magnitude higher than the movement through a clay layer. However, it is just as important to clean up the clay layer as it is the sand layer. Determining when to stop is an important consideration in developing a remediation program.

One of the problems in working with bioremediation is determining the end point, or determining when the system is fully bioremediated. One theory insists that the bioremediation is complete when the monitoring wells around the periphery of the plume show no traces of hydrocarbons. Use of this theory can cause premature cessation of remediation activities, especially where clays are involved. If a clay is present and has an amount of contamination, the nutrients that support the bacteria must work their way in from the outside, and the process is reaction-rate limited. Any hydrocarbons escaping into a zone where sufficient oxygen and nutrients exist will be diluted and degraded rapidly.

Another theory relies on the measurement of bacterial activity through measurement of oxygen and carbon dioxide levels in the monitoring wells around the site. When the oxygen concentration approaches a stable value for a reasonable time, there is a likelihood that all biodegradation has ceased and the well is reaching equilibrium with the injection well. Also, if the CO_2 levels and total organic carbon levels in the monitoring wells at the perimeter of the plume decline and remain low for a reasonable period of time, one may conclude that there is no further bioremediation taking place because there is no carbon dioxide being produced. The total organic carbon test provides additional confirmation of the completion of the biomonitoring activity.

A final confirmation of the completion of bioremediation is the sampling of the soil in the middle of the most contaminated and most resistant zones. If the soil is found to be clean and there is no contamination present, the site should be decommissioned.

3.10.7 TAKING THE MEASURE OF BIOREMEDIATION

The effectiveness of bioremediation has been estimated but never effectively proven. Bioremediated sites are not supposed to have any residual contamination, but few sites are monitored for a long enough period to insure that the bioremediation effort is effective. Bioremediation activities require time and money. The real savings from bioremediation are seen by comparison to other alternatives that could require the formation to be pumped for many years. The initial costs for a bioremediation system are somewhat higher than those of a pump-and-treat system, but the total costs are substantially lower.

For all the talk of being faster and better, bioremediation is a process whose progress may be measured in months and years. Many of the compounds in gasolines are somewhat bioresistant, and the effective half-lives of these compounds are measured in months. The microbial degradation rates are measured in micrograms of material decomposed per gram of active culture per hour or per day. When the half-life of a compound is one month, it will require one month to remove 50%, two months to remove 75%, and so on. If an area has a contaminant level of 100,000 parts per million in the soil (10% by weight), and the half-life of the degradation of the compound is one month, by the end of one year, there will still be approximately 24.4 parts per million left if everything goes as planned. By comparison, a pump-and-treat system may require several years to achieve the same levels if the material can be washed out of the soil.

Bioremediation is an aggressive remediation technique in that it requires a lot of pumping and a lot of treating of the recovered plume and the chemicals, and none of this activity is cheap. Bioremediation will only work on the organic contaminants in the soil; the process will not remove lead or other inorganic compounds present. If lead levels in the soil are high, they may inhibit bioremediation. By the time one is finished bioremediating the soils at a site, several pore volumes of soil water will have been replaced.

Bioremediation systems work best in sands and gravels, the same type of soils in which other remediation systems also work well. Bioremediation does have the advantage that it can remove and degrade the organic compounds while the other systems cannot. Bioremediation as a treatment following vapor stripping holds much promise for cost-effective remediation.

3.11 SOLIDIFICATION AND CONTAINMENT

Solidification and containment are really not treatment systems but are passive measures and preventive measures to be taken while something else is happening. The acknowledged exception to this is vitrification, which will be discussed later.

Solidification is a process in which the soil is incorporated in a matrix that hardens or cures with time. One of the purposes of the process is to reduce or eliminate potential problems with leaching from the soil.

Containment is a passive process as well in that it relies on exterior elements to hold the soil and the contamination in place for a while until other things happen.

3.11.1 CONTAINMENT

Containment is a process of building a wall around a contaminant. The wall may be clay, a clay slurry (slurry wall), concrete, sheet piles, or almost anything else. Containment works best when there is an unweathered rock layer, clay layer, or other aquaclude below the contaminant area. Unbroken rock is often preferred because many petroleum hydrocarbons have non-polar molecular arrangements that will cause desiccation and cracking of clays. These cracks allow the chemicals to pass through the clay almost as if it were not there. Experiments conducted several years ago by the U.S.EPA's Cincinnati research laboratory indicated that even in the presence of water, non-polar organic chemicals could attack and desiccate a clay, and could increase the permeability of a very impermeable clay to values comparable with those found in sands and silty sands.

Containment can also be used to prevent further vertical migration of the contaminants. At a typical service station site, the pavement often serves as a good cap to contain the volatile organic materials in the area and to prevent rainfall from washing the contaminants further down in the formation.

A few years ago the U.S.EPA attempted to develop a system of bottom sealing a site to prevent vertical migration of chemicals from underneath. The selected technology was jetting the well in combination with a chemical fixative or soil-cement. The idea was a bust; the number of wells required made the costs prohibitive, and the integrity of the subsurface seal was marginal to poor. The importance of having a bottom seal or containment layer of unfractured unweathered rock to which the containment system can be anchored cannot be too strongly stressed.

3.11.2 SOLIDIFICATION

Solidification is the process of making contaminated soil into a solid. Mixing soil with a cement matrix or with a sodium silicate matrix will accomplish the solidification. The problem for petroleum-contaminated sites is that the materials do not do a very good job of controlling the leachate from the fixed soil matrix.

The technology is reminiscent of alchemy because it has many arts and recipes, and each is closely held and protected with secrecy agreements. Despite the manufacturer's claims of the ability to process and stabilize organic chemicals (and some even claim to be able to prevent volatilization), the solidification systems work best on inorganic contaminants such as metals. When one discusses solidification, it is best to remember that the people promoting a specific process do not mention that their systems work best when they handle no more than 50% of contaminated material in their matrix, to achieve a good solid material. Petroleum hydrocarbons are not compatible with cement products or with sodium silicate products. Hydrocarbons will retard the set of the material and can stop the solidification process altogether.

In order to accomplish solidification, the soil must be excavated, stockpiled, mixed with the solidification chemicals, and allowed to cure. At the end of the solidification process, there is at least 30% more material by volume than there was at the start of the process.

One of the few solidification processes that works effectively on petroleum-contaminated soils is asphalting. The technology has been used in the nuclear industry for treating radioactive materials, and its effectiveness has been proven. It is expensive, and the heat in the process will evaporate many of the compounds. It also requires excavation and handling of the soil.

The new TCLP test that the EPA uses to determine whether or not a waste is hazardous incorporates many more compounds than the EP toxicity test did. Under the new test procedures, (located in 40 CFR 261), as little as 0.5 mg/l of benzene in the waste extract from the TCLP test can cause a soil to be listed as a hazardous waste. It is possible that some solidification processes may be able to seal the soil effectively, but the long-term viability of the process has not yet been determined.

3.11.3 VITRIFICATION

Vitrification is one of the few solidification technologies that is considered effective under a variety of circumstances. Batelle Industries have developed the process and are promoting the technology for sites that are hard to treat in other ways.

In the vitrification process, a relatively small area, no greater than about 20′ by 20′, is selected for processing. Carbon electrodes are sunk to a level below the contaminant. High-voltage current is applied between the electrodes, and the resistance of the soil generates heat that ultimately melts the soil and converts it to glass. From the literature and presentations on the subject, the process is effective, but moderately expensive. It may be more expensive than the cost of incineration.

Air pollutants from the process are controlled by erecting a tent over the vitrification site and by using portable air pollution control equipment that is brought to the site with the electrodes and electrical generating equipment. The process may be effective in treating the residuals left on a petroleum-contaminated site, but the heat generated by the process would volatilize and remove the hydrocarbons from the soil, rendering the treatment ineffective. A demonstration of the vitrification process reportedly affected a relatively small area outside the treatment area. The soil around the outside of the treatment area has an elevated temperature for only 5-8 feet around the outside of the site. This is relatively small when one considers that the temperature within the heated soil can be in excess of 1500°. This technology may be inappropriate for treatment of petroleum-contaminated soils at an active service station or terminal site because of the high temperatures and the potential for fire. Despite Battelle's claims that the process is completely developed, the cost information is tentative at best.

Batelle has attempted to promote the technology for stabilizing nuclear contaminated soils at the Hanford Nuclear Test Center near Richland, Washington. In a 1990 test, the containment canopy over the solidification site caught fire and caused a test site failure. Batelle may be considering abandonment of this technology.

3.12 SUMMARY OF REMEDIATION TECHNOLOGIES

Many of the remediation technologies considered in this chapter required extensive excavation and handling of the soil. Incineration, surface biotreatment, excavation, and solidification all fall into this category.

The areas that hold the most promise for the most cost effective treatment of contaminated soils are vapor stripping, followed by *in-situ* bio-treatment. Trenches, wells, ditches, etc. are all valid treatment techniques if they serve the purpose of controlling the migration of the contaminants at the site. In the final analysis, a successful remediation effort depends upon the site geology, and the remediation engineer must make his choices to optimize remediation effort with respect to the geology at the site. Anything less is expensive and ineffective.

REFERENCES

1 U.S.EPA, Office of Research and Development. *Handbook–Remedial Action at Waste Disposal Sites (Revised)*. Hazardous Waste Engineering Lab. Publication Number EPA/625/6-85/006.

2 Russell, D. L. 1987. "Understanding Groundwater Monitoring," *Chemical Engineering Magazine*, October 28,

3 Hardenburg and Rodie. *Water and Wastewater Engineering*. New York: McGraw Hill.

4 Nyar, E. K. *Ground Water Monitoring Review*.

5 Reynolds, T. D. 1982. *Unit Operations and Processes in Environmental Engineering*. Belmont, CA: PWS Publishers, p. 187-214.

6 Envirite Field Services. November 1989, Contract Proposal for Bog Creek Farm Superfund Project.

7 Personal correspondence with Mr. Mark Shearon, President of Harmon Environmental Services.

8 U.S.EPA, Office of Underground Storage Tanks. 1989. *Estimating Air Emissions from Petroleum UST Cleanups*. Washington, D.C.

9 U.S.EPA, Office of Underground Storage Tanks, op. cit.

10 Mutch, R. D., Jr., A. N. Clarke, J. H. Clarke and D. J. Wilson. "*In situ* Vapor Stripping: Preliminary Results of a Field-Scale U.S.EPA/Industry-Funded Research Project," *Proceedings, Industrial Waste Symposium, Water Pollution Control Federation Annual Meeting, San Francisco, CA, 1989.*

11 Sims, R. D., op. cit.

12 Sims, R. D., op. cit.

13 Oberdorfer, J. A. and F. L. Peterson. 1985. "Waste-Water Injection: Geochemical and Biogeochemical Clogging Processes," *Ground Water Monitoring Review*.

CHAPTER 4

Costs of Remedial Activities

4.1 INTRODUCTION

CLEANUP of gasoline-contaminated service stations is very expensive. In some instances, the cleanup is more expensive than it needs to be. This manual and this chapter have attempted to address cost containment and management of remedial activities. In this chapter the costs of remedial services and equipment are presented and discussed. The first section in this chapter deals with the general topics of cost and the management of various aspects of remediations. The second section of this chapter deals with specific remediation equipment costs and associated construction costs as far as they can be determined. The appendices to this chapter present unit cost information from a chemical laboratory and from a geology/engineering and exploration firm. The third section presents an abstract of theoretical cost projections for equipment installation and operation from recent American Petroleum Institute publications.

4.1.1 INFORMATION SOURCES

The unit construction cost information in this chapter is based upon the information contained in the nationally recognized *R.S. Means Building Construction Cost Data—1990*. Manufacturer's quotations were obtained and included where appropriate for the purposes of estimation. An attempt was made to relate the estimates prepared from *Means* to the U.S.EPA's cost information from their 1985 report: *Compendium of Costs of Remedial Technologies at Hazardous Waste Sites*, but the EPA data was incomplete, and of marginal value.

Typical equipment rental rates and laboratory analysis costs are shown in the appendix to this chapter. The equipment rental rates were provided by Versar, Inc., and the laboratory costs were provided by Analytical Services Inc., Atlanta. Many geologic consulting and exploration firms are reticent to discuss their charges and unit rates. Others are more open about the information. No attempt was made to relate the rate structures of any drilling or investigation firm to project costs because of the inherent differences in each site. The cost data in Appendix A contains hourly rates and fees information from Atlanta Testing and Engineering, Duluth, Georgia. Their rates are representative of typical drilling and geo-technical exploration rates prevalent throughout the Southeast.

The drilling rates schedule information, labor rates, laboratory charges, and equipment rental information in this chapter is not to be construed as indicative of the quality of the work or the value of the service performed or provided, as those are value judgments that must be made based on specific factors. The cost information provided is representative of the fees and services unit rates and fixed prices charged by responsible professionals in the Southeast.

Appendix B contains information reproduced from various sources made available from the American Petroleum Institute. The information was generated from theoretical modeling studies, by different subcontractors. The important tables and estimated cost data for hypothetical sites are reproduced from each of the studies in Appendix B [1,2].

4.1.2 DISCUSSION

The remediation contractor is allowed to investigate the site, design the remediation system, install the equipment, and maintain and operate the equipment. The contractor operates as the engineer, general contractor, and the maintenance contractor. The remediation contractors' markups are substantially above those used for other engineering/construction companies in comparable industries. While the initial justification for the higher markups was the additional liability and costs associated with doing work on a hazardous waste site, changes in companies' perspectives on liability, changes in the hazardous substance and hazardous waste regulations, and the imposition of OSHA rules regarding personnel protection has forced many of the "classical" engineering firms to reconsider their participation in remediation projects. The entrance of these large engineering houses will substantially change the shape of the marketplace in the next few years.

The design and construction engineering business is very efficient for projects where cost factors can be determined in advance, and where a budget and a detailed scope of services can be prepared in advance of performing work on a project. In these instances, the contractors have an incentive to carefully determine and manage their costs to bring the project in under the client's budget. The advantage in using these firms is that they have a variety of disciplines, including construction and purchasing management services available from one source.

When an owner or operator of a site establishes a budget and a schedule for a remediation project, the project can easily be monitored and managed to determine its performance against the budget. With a very complete site characterization study, and the development of key information as outlined in previous chapters, an entire remediation project budget can be established and bid upon by independent contractors. Design/construction firms and large construction firms are hungry for this type of work.

By contrast, there seems to be too much money available in the UST cleanup business and too few resources available to permit client companies to adequately manage the large number of UST sites and contractors. The lack of management resources and contractor self-interest has led to a number of cost overruns on projects.

One way in which large contractors can reduce remediation equipment costs is by fuller utilization of national purchasing power. The number of equipment suppliers is extremely small, and the markup on the equipment is high with respect to the manufacturing cost. The savings from using the discounts available under a national contract might be on the order of 20% of the list price, and as much as 40% when the contractor's markups are considered. Site owners should also consider development of specifications and standards that would enable the projects to be competitively bid to a fixed price.

4.2 COST FACTORS—DISCUSSION

4.2.1 EXPLORATION SERVICES

In every exploration project there are three elements of cost – equipment, labor, and analyses. The equipment costs represent the annual operating and replacement costs of the equipment plus a profit. The labor costs are marked up by overhead factors between 100% and 250% to cover the benefits, medical monitoring, office expenses, and other items associated with the work. Fee and rate structures are frequently based upon category rates that express a common hourly rate charge for various employees with common work classifications but slightly differing levels of experience. The alternative to hourly rate schedules is a direct multiple of salary. Drilling and equipment charges are generally based upon a mobilization fee plus a charge per foot of well drilled, depending upon the diameter of the well.

4.2.2 LABORATORY COSTS

The cost of laboratory services can vary widely depending upon the skill of the laboratory, the methods of analysis, and the qualifications of the laboratory. As a general rule, the more qualifications and equipment the laboratory has, the greater its overhead and prices. A laboratory that participates in the Contract Laboratory Program (CLP) or that is a "state-certified" laboratory, generally has a higher overhead than one that does not. CLP and state-certified labs are required to perform a number of duplicate equipment calibration analyses for each sample analyzed. The laboratory Quality Assurance and Quality Control Program is designed to produce highly accurate results (generally within plus or minus 1%). The commercial non-CLP or non-certified laboratory may be nearly as accurate, but doesn't have the constraints of the CLP program and as such can provide the work at a much cheaper price. However, the state does not often accept data from non-certified laboratories without much scrutiny. Where the project does not require highly accurate analyses, such as in the initial stages of an investigation, substantial savings can be obtained by using a non-CLP laboratory.

Depending upon the remediation contractor and the terms of the contract, the contractor will receive a discount on the laboratory services he sends to a subcontract laboratory. That discount may be between 10 and 20%, depending upon the volume of work generated for that laboratory. Depending upon the good intentions of the contractor, the reduced costs may or may not be passed along to the client or owner, as the discount for the laboratory services is generally granted at the time of billing and may be by separate invoice.

If the owner or operator of a site selects regional laboratories and develops a master contract with those laboratories, he or she will receive a discount that could amount to between 10 and 25% of the cost of laboratory services. A 10-15% discount is available just for the asking. If the volume of work is large (on the order of several thousand samples per year), the discount for using the laboratory on a semi-exclusive basis will be closer to 25%.

Negotiating master contracts has a secondary advantage: priority service. Often, a laboratory takes up to four weeks to perform the analysis of a single sample by GC/MS. By taking advantage of the purchasing power of the large number of analyses to be run, the owner or operator can obtain priority service at little or no additional cost.

4.2.3 SITE INVESTIGATION COSTS

The site investigation and exploration specifications currently used by several major gasoline retailers call for the drilling of five holes at a depth of up to 25 feet for a typical installation, the logging of the borings, and the preparation of a report. For a typical installation, the costs for the work are generally between $8000 and $10,000 per site. Sailors Engineering Associates, a geo-technical, testing, and remediation contractor, estimates that the installation of a 2″ PVC monitoring well will cost about $30 per vertical foot, and a 4″ diameter recovery well will cost about $45 per vertical foot.

These costs may be representative throughout the industry. On a site basis, the drilling costs would represent between $4000 and $6000. Atlanta Testing and Engineering suggested that another way to reduce the exploration costs is to instruct the driller to cut off the borings at 20 feet if the boring above that depth showed no evidence of contamination. The boring costs for exploration at these sites has generally been in the range of $6000-$8000 per site, including the report.

Contractors can be asked to prepare bids to show unit rates and man-hour estimates, markups, and other charges, and they do so on a regular basis for many of their clients. Contractors are willing to work on a cost plus basis to an upset price so that the total project costs are controlled. As more information about the unit prices and cost is requested, the remediation contractors will become aware of increased contract scrutiny, and the pressures of competition will reduce costs on the various projects. The principal difficulty with this system is that it requires management time and attention, which will burden already busy staff.

The competitive bidding procedure can work to a site owner's economic advantage. By competitively bidding initial investigations, and by carefully examining cost factors that may apply to continuing work by a single contractor on a second phase of work, an owner can reduce overall exploration and investigation costs.

The selection of construction materials for a well can greatly affect the total cost of the project. PVC will undergo some slight degradation by swelling and softening when immersed in gasoline, but that is tolerable when the life of the project is relatively short, say five years or less. Most of the remediations do not involve the investigation for metals in the groundwater. Galvanized steel, or plain steel pipe can be used for the well screen, with satisfactory results.

When stainless steel is used for a screen material it has an indefinite life. Owners often neglect the possibility of re-using the screen. For a well screen above 6 inches in diameter, it might be cheaper in the long run to use a PVC or galvanized screen, accept the deterioration, and replace the well or the casing when the need arises. By these means, a well could have a service life that approaches twelve years at a lower total cost than initial installation of the stainless steel screen.

4.2.4 DESIGN, CONSTRUCTION, AND MAINTENANCE COSTS FOR REMEDIAL ACTIVITIES

4.2.4.1 Construction Cost Control

Once contamination has been discovered at a retail outlet it must be treated. The most common ways of treating the soils and waters are discussed in Chapter 3. Each treatment technology has a certain type of cost associated with it. The selection of a particular treatment technology *without* complete analysis of the total cost of the project over its duration can lead to erroneous decisions regarding the selection of the most cost-effective technology.

Many "full service" contractors have a vested interest in the installation and operation of treatment equipment. When the contractor is marking up the cost of the equipment by 15% from list price, the client is not receiving the cost benefit that can be gained through competitive bidding. The contractor has no incentive to require the manufacturer to reduce his price or give any discount from list price. Often, there is no national contract with equipment suppliers, and each contractor is free to select equipment from the vendors on any basis. The result is that the client winds up with a hodgepodge of different types of equipment at each site. When a site owner changes remediation investigation contractors, the new contractor often wants to change all the equipment on the site, often for no apparent reason.

4.2.4.2 Cost-Effective Design

One of the ways in which the equipment and service costs can be lowered is through proper use of instrumentation. The use of simple design features and proper instrumentation can even summon a service technician in the event of a mechanical breakdown. In the event of a mechanical equipment failure, the monitoring equipment signals the auto-dialer, which calls a predetermined number and plays a recorded message.

Construction costs can also be lowered through equipment standardization. Selection of a few manufacturers of remediation equipment and development of standard design specifications can create substantial savings for the owner or operator responsible for paying for a cleanup.

The remedial technology gaining most wide acceptance at gasoline-contaminated sites utilizes vacuum stripping of the soil, with or without incineration of the vented vapors, and utilizes stripping towers to remove gasoline from the groundwater. The use of standard equipment packages and designs available from manufacturers directly would greatly reduce the cost of equipment installation and design.

Many manufacturers make equipment in specific design sizes, usually 1, 5, 10, and 25 gallon-per-minute units. With standard height and dimensions, the installation contractor does not design a stripping tower, but provides the equipment manufacturer with a water analysis, and specifies the treatment level to be obtained from the tower. The equipment manufacturer has a number of standard designs that he quotes to the contractor. If a contractor deals directly with an equipment supplier and obtains a national or regional contract, substantial savings can be realized. The equipment that should be standardized includes pumps, blowers, carbon adsorption units, stripping towers, demisters, vapor combustors, and fume incinerators.

4.2.4.3 Maintenance Cost Controls

Equipment maintenance is always good for a contractor's revenues, as it will allow the contractor to provide steady employment for an employee, even if the employee does little but run to each remediation site periodically to perform monitoring and maintenance work – work that can often be better performed by others at a lower cost.

The personnel sent out to check the mechanical operation of a remediation system are not qualified mechanics, but sampling technicians, draftsmen, or driller's helpers. Most of these people have little specialized knowledge about the remediation system, its operation, or repair. The principal qualifications of these technicians are that they received the OSHA-required 40-hour training. If these technicians determine that a pump is not running, they do not fix it, but report the equipment failure or ask someone else to call in a qualified serviceman. If one were to judge the quality of the equipment maintenance and repair from some of the service witnessed in the field, it would be possible to conclude that it might be best to call in Sears Home Appliance Repair Service. Sears, at least, sends out a technician/mechanic who has received some factory training and can at least make the repairs or who knows how and where to get the needed spare parts quickly.

As an alternative to the high cost of providing service technicians, a number of firms can be selected to service and maintain the equipment at a much lower hourly rate and total cost. Service technician rates for plumbers and air conditioning repairmen can be used as a guide to comparing hourly rate and cost information for service contracts.

The trade unions are starting to provide members with HAZWASTE 40-hour training so that they can legally work on projects where remediation is being performed. One particular advantage of the use of union labor is that the laborers and journeyman-grade tradesmen have to meet certain minimum certification skills before they join the union and obtain a journeyman's license.

4.3 SPECIFIC COST FACTORS

This section contains specific equipment pricing estimates gathered from a number of sources. Wherever possible, the sources are identified. While this section can provide useful information on what some things cost, it cannot determine what a project will cost.

It is difficult and misleading to determine in advance, and on a general basis, the labor component of the various scientific contractors for a specific site. Their relative efficiency, the number of man-hours, the cost of their labor, the experience of their labor, and the size and complexity of a typical project will determine what their final charges will be. There are too many uncertainties in any given project to enable accurate determination of the total project costs by someone removed from the project.

4.3.1 EQUIPMENT COST FACTORS

The unit cost factors in Table 4.1 are for comparable equipment, without installation labor, unless otherwise indicated. Contractor's standard markups or list prices are used in determining equipment costs.

TABLE 4.1. Equipment Cost Factors.

Equipment Item	Manufacturer	Approximate Cost
Well pumps		
Pneumatic pumps	R.E. Wright	$ 1700 per pump
Pneumatic controller	R.E. Wright	$ 4000 two pumps $ 7000 six pumps
Pneumatic pump and controller	Ejector Systems	$ 7550 for 1 well $ 3270 each add'l well
Pneumatic pump and controller	Westinghouse Groundwater Recovery	$ 2100 per pump
Air compressor for pumps—50 psig continuous 5 cfm		$ 1800
Electric pump no controls	Grundfos	$ 1300 per pump depending upon size
Stripping towers—no water treatment		
Stripping towers 1′ diam. 3′ diam.	R.E. Wright	$ 1250/month—lease $ 2250/month—lease
Low profile tower—no controls 2-3 gpm	Ejector Systems	$ 8000
Low profile tower—full control 35 gpm	Ejector Systems	$33,000
Stripping tower—package unit —stainless steel 1′ diam. 2′ diam.	Westinghouse	$ 6500 $12,500
Demister for stripping tower		$ 300-900
Oil water separators		
Modular, 10 gpm Modular, 40 gpm (includes tank and controls)	Westinghouse G.R.	$11,500 $14,200

TABLE 4.1. (continued).

Equipment Item	Manufacturer	Approximate Cost
Separator only, 7 gpm	Ejector Systems	$ 5000
Separator only, 35 gpm		$ 9400
Separator and filter cartridge style, 30 gpm	Serfilco	$ 3200
Carbon adsorbers—water systems		
55 gallon canister, 110 carbon charge	Tigg	$ 460
300 lb carbon canister	Tigg	$ 1360
3000 lb carbon canister	Tigg	$ 6575
Vapor recovery exhaust pumps—pumps only		
150 cfm @ 4″ Hg	Lamanson	$ 2800
250 cfm @ 6″ Hg		$ 3200
Fume incinerators		
150 cfm catalytic combustor	Anguill Environmental	$45,000*
150 cfm vapor incinerator	Nat'l Air Oil	$41,000*
150 cfm vapor incinerator	McGill Environmental	$79,000
Carbon adsorber—air system		
100 cfm, 200 lb carbon charge	Tigg	$ 710
500 cfm, 1900 lb carbon charge	Tigg	$ 7050
Remedial investigation services		
See information in Appendix A regarding typical investigation services costs from Atlanta Testing and Engineering.		
Analytical services		
See typical pricing from Analytical Services and from Savannah Laboratories in Appendix A.		
Specialty equipment for remedial investigations		
See typical pricing on equipment rentals from Versar, Inc., in Appendix A.		
Incineration		
Cost per ton incinerated	Various contractors	$50-90**

*Quoted without water trap. Water trap adds $7500 to catalytic combustor and approximately $4000 to National Air Oil incinerator.

** Exclusive of excavation.

TABLE 4.2. Unit Pricing for Remedial Work.

Pavement removal and replacement	
Asphalt	
remove	$ 5.60/sq.yd.
replace	$ 10.30/sq.yd.
Concrete, 8″ thick	
remove	$ 6.20-11.75/sq.yd.
replace	$ 19.25/sq.yd.
Excavation and disposal	
Site excavation—backhoe	$ 2.00/c.y.
Waste soil disposal—non-hazardous landfill	$ 75-125 per ton
Clean soil backfill and compact	$ 9.00-15.00/c.y.
Trenching and underground piping	
4′ deep × 3′ wide, no bracing	$ 1.50/l.f.
8′ deep × 4′ wide with bracing	$ 12-30/l.f.
Recovery trench, gravel backfill	$ 35.00/l.f.
add for pipe laid in trench	
4″ PVC laid in trench	$ 2.60/l.f.
6″ PVC laid in trench	$ 3.40/l.f.
add for piling if required below 5 feet	
Wood sheet piles-sheeting if req'd in trench—8′ deep	$ 55.00/l.f.
Manholes and well box covers	
Well box w/locking cover	$ 70.00 ea.
4′ diam. manhole, with light duty frame and cover and base in place, approximately 4′ deep	$ 855.00 ea.
4′ diam. manhole same as above, 8′ deep	$1090.00 ea.
Add for each vertical foot over 8′	$ 150.00/v.l.f.

TABLE 4.2. (continued).

Miscellaneous construction cost items	
Pre-engineered buildings—erected	$ 15.00/s.f.
Electrical panel/motor control center	$1500.00 ea.
Electrical motor starters	$ 130-500 each depending upon size
Electrical wiring, direct burial armored cable	$ 3.00/l.f.
Electrical wiring, pulled through conduit	$ 2.30/l.f.
Electrical connections to motors	$ 120.00 ea.
Product recovery tanks—aboveground 275 gal	$ 375.00 ea.
Product recovery tanks—aboveground 550 gal	$ 700.00 ea.
Fencing—chain-link 6′ high plus 3-strand barbed wire	$ 20.00/l.f.
Typical labor rates, national average for skilled trades—union	
Labor rates include all fringes and benefits, and contractor's profits.	
Average 35 trades	$ 33.65/hr.
Carpenters	$ 30.60/hr.
Cement finishers	$ 30.60/hr.
Electricians	$ 35.85/hr.
Equipment operators—backhoe	$ 34.20/hr.
Laborers	$ 26.05/hr.
Plumbers	$ 36.65/hr.

4.3.2 UNIT CONSTRUCTION COST AND LABOR RATE INFORMATION

Table 4.2 addresses the labor costs for a number of remedial alternatives where cost factors could be readily identified.

REFERENCES

1 API. 1988. *Treatment System for the Reduction of Aromatic Hydrocarbons and Ether Concentrations in Groundwater*. API Publication Number 4471.

2 API. 1966. *Cost Model for Selected Technologies for Removal of Gasoline Components from Groundwater*. API Publication Number 4422.

APPENDIX A

Specific Pricing Information

GEOLOGIC AND FIELD SERVICES

GEOLOGIC SERVICES

Geologic technician, per hour	$ 35.00
Associate geologist/hydrogeologist, per hour	$ 45.00
Staff geologist/hydrogeologist, per hour	$ 55.00
Project geologist/hydrogeologist	$ 65.00
Senior geologist/hydrogeologist, per hour	$ 75.00
Chief geologist/hydrogeologist or consultant, per hour	$ 95.00
For time spent in depositions and/or court appearances, –multiply above rates by 1.5	
Secretarial and drafting services for report preparation, per hour	$ 25.00
Report reproduction (initial three copies at no charge), per page	$ 0.25
Special printing or blueprinting charges–actual cost plus 25%	
Mileage, per mile	$ 0.30
Travel expenses and other reimbursables–actual cost plus 25%	
Long distance telephone calls–actual cost	
Computer services charge, per analysis	$ 200.00
Seismic exploration, per hour	$ 100.00
Resistivity survey, per hour	$ 100.00
Spontaneous potential survey, per hour	$ 100.00
Terrain conductivity meter (equipment charge), per day	$ 200.00
per week	$ 500.00
Soil-gas radon survey (equipment charge), per day	$ 150.00
Indoor radon measurement by granular activated carbon canister, per canister	$ 30.00
Soil gas analysis for VOC (equipment charge), per day	$ 150.00
Other surface geophysical surveys (rates available on request)	
Borehole logging (rates available on request)	

ENVIRONMENTAL EXPLORATION

Mobilization of drilling equipment to the site (includes rig, two-man crew, and initial off-site steam cleaning)	
In-town, lump sum	$ 350.00
Out-of-town – $5.00 per mile one way plus	$ 100.00
Soil Test Boring[3]	
Soils with penetration resistances less than 50 blows per foot, per linear foot, sampling at 5-foot intervals[4]	
3¼-inch I.D. hollow-stem auger	$ 8.00
6¼-inch I.D. hollow-stem auger	$ 11.00
Wet rotary – up to 6-inch diameter	$ 9.00
Wet rotary – large diameter (scope dependent)	
Soils with penetration resistances greater than 50 blows per foot, per linear foot, sampling at 5-foot intervals[5]	
3¼-inch I.D. hollow-stem auger	$ 10.00
6¼-inch I.D. hollow-stem auger	$ 13.00
Wet rotary – up to 6-inch diameter	$ 11.00
Wet rotary – large diameter (scope dependent)	
Auger boring[6]/wet rotary (no sampling), per linear foot	
3⅓-inch I.D. hollow-stem auger	$ 5.00
6⅓-inch I.D. hollow-stem auger	$ 8.00
Wet rotary – up to 6-inch diameter	$ 6.00
Wet rotary – large diameter (scope dependent)	
Hourly services (rig and two-man crew)	
May include drilling, well installation, decontamination, well development, difficult moving, and delay (not related to weather or mechanical breakdown), per hour	$ 115.00
Additional crew member, per hour	$ 25.00
Additional crew member with pickup truck, per hour	$ 35.00
Equipment	
Steam cleaner, per day	$ 60.00
Self-contained steam cleaning unit, per day	$ 110.00
Grout unit, per day	$ 60.00
Water trailer (500 gal.), per day	$ 50.00
Water truck (1000 gal.) with driver, per hour	$ 40.00
Water truck (1000 gal.) without driver, per day	$ 150.00
Materials	
Well materials, decontamination supplies, health and safety supplies – actual cost + 25%	

[3]Depths less than 60 feet. For depths greater than 60 feet, add $2.40 per linear foot.
[4]Extra split spoon sample, $18.00 each.
[5]See footnote 4.
[6]See footnote 3.

Out-of-town expenses

Two-man crew, per day	$ 100.00
Three-man crew, per day	$ 140.00

Health and safety surcharges

Level D – add 10%
Level C – add 50%

LABORATORY SERVICES

PRICING SCHEDULE: WATER AND WASTEWATER ANALYSES[7]

Parameter	Price
Acidity	$ 5.00
Alkalinity (phenolphthalein)	10.00
Alkalinity (total)	10.00
Aluminum	15.00
Ammoniacal nitrogen (distillation)	15.00
Antimony	20.00
Arsenic	20.00
Barium	15.00
Beryllium	20.00
Bismuth	25.00
Biochemical oxygen demand (5-day)	20.00
Boron	17.00
Bromide	35.00
Cadmium	15.00
Calorimetric analysis (BTU)	30.00
Calorimetric analysis (halogen and sulfur)	40.00
Calcium	15.00
Carbon, total organic	37.50
Carbon dioxide, free	no charge
Chloride	10.00
Chlorine demand	50.00
Chlorine residual	10.00
Chromium, total	15.00
Chromium, hexavalent (in water)	12.00
Chemical oxygen demand	20.00
Cobalt	15.00
Color (Pt-Co units)	10.00
Conductivity	10.00
Copper	15.00

[7]Other analyses available upon request.

Parameter	Price
Cyanide, amenable to chlorination	$ 40.00
Cyanide, total (distillation)	25.00
Fecal coliform	20.00
Formaldehyde (chromotropic acid)	35.00
Fluoride	15.00
Hardness, total (titration)	10.00
Iron	15.00
Lead	15.00
Magnesium	15.00
Manganese	15.00
Methylene blue active substance (MBAS)	25.00
Molybdenum	25.00
Nickel	15.00
Nitrogen, ammoniacal (distillation)	15.00
Nitrogen, kjeldahl	20.00
Nitrogen, nitrate	15.00
Nitrogen, nitrite	15.00
Nitrogen, organic	20.00
Oil and grease	25.00
Organic carbon, total	37.50
Oxygen, dissolved	5.00
pH	5.00
Phenols, total	25.00
Phosphate, total	18.00
Phosphate, ortho	15.00
Potassium	15.00
Selenium	20.00
Silica, dissolved	15.00
Silver	15.00
Sodium	15.00
Solids	
Total	10.00
Total dissolved	10.00
Total suspended	10.00
Volatile suspended	10.00
Total volatile	10.00
Settleable	10.00
Strontium	20.00
Sulfate	15.00

Parameter	Price
Sulfide	$ 17.00
Sulfite	17.00
Surfactants	see MBAS
Tannin and Lignin	25.00
Tin	20.00
Titanium	20.00
Thallium	20.00
Turbidity	10.00
Vanadium	20.00
Zinc	15.00

PRICING SCHEDULE: RCRA TESTING

Parameter	Price
Ignitability	
Flash point (Pensky-Martens)	$ 25.00
Flash point (TCC)	25.00
Corrosivity	
pH	8.00
NACE standard TM-01-09	50.00
Reactivity	
Cyanide	35.00
Sulfide	30.00
Sulfite	30.00
EP Toxicity	
Extraction procedure	50.00
Oily waste extraction	150.00
Extraction procedure plus:	
8 metals (As, Ba, Cd, Pb, Hg, Se, Ag)[8]	150.00
Pesticides and herbicides (Endrin, Lindane, Methoxychlor, Toxaphene, 2,4-D, 2,4,5-TP Silvex)	200.00
Partial RCRA scan	
Includes Ignitability, Corrosivity, Reactivity, Extraction, and Analysis of 8 Metals[9]	225.00
Complete RCRA Scan	
Includes above parameters plus pesticides and herbicides[10]	300.00

All tests are performed in accordance with *Test Methods For Evaluating Solid Waste*, EPA SW-846, 1986.

[8]For method of Standard Additions (delisting procedure), add $100.00.

[9]See footnote 8.

[10]See footnote 8.

Parameter	Price
Toxicity characteristic leaching procedure	$1200.00
Hazardous waste profile	quote

PRICING SCHEDULE: DRINKING WATER

Community Water Systems Group Analyses		$ 290.00
Includes:		
Aresenic	Chloride	
Barium	Copper	
Cadmium	Iron	
Chromium	Manganese	
Fluoride	Sodium	
Lead	Sulfate	
Mercury	Total dissolved solids	
Nitrate	Zinc	
Selenium	Color	
Silver	Turbidity	
Alkalinity	Carbon dioxide	
Hardness	pH	
Nitrate	Turbidity	
Chloride	Alkalinity	
Iron	Hardness	
Manganese	Carbon dioxide	
Color	pH	
Individual wells (bank closing approval)		$ 25.00
Bacteria (fecal coliform)		
Trihalomethanes		$ 130.00

PRICING SCHEDULE: PRIORITY POLLUTANTS

Parameter	Price
Volatiles (VOC)(Method 624, GC/MS)	$ 185.00
Acid extractables (Method 625, GC/MS)	200.00
Base-neutral extractables (Method 625, GC/MS)	235.00
Pesticides (Method 608, ECD)	90.00
PCB's (Method 608, ECD)	90.00
13 metals	150.00
Total cyanide	25.00
Total phenols	25.00
Total toxic organics (TTO)	800.00
127 priority pollutants	950.00
Dioxin	quote
Asbestos	quote

MONITORING SERVICES DEPARTMENT–EQUIPMENT LEASE PRICE LIST

Equipment Category	Description	Daily	Weekly	Monthly
Air	Air Velocity Gauge with 18″ Pitot Tube	$ 4.50	$ 22.52	$ 45.03
Air	Anemometer, Recording	17.48	87.40	174.80
Air	Calibration Kit for Hi-Vols	3.21	16.06	32.11
Air	Calibrator, Pump, (300-3000)	3.93	19.67	39.33
Air	Calibrator, Pump, (500-6000)	4.40	21.99	43.99
Air	Cascade Impacter, Five Stage	20.43	102.13	204.25
Air	Chart Recorder w/Alarm 1 Day Paper, Circular	22.80	114.00	228.00
Air	Detector Pump, Drager	4.75	23.75	47.50
Air	Dry Gas Meter, NBS Traceable	44.46	222.30	444.60
Air	Flow Cell Assembly, Standard	7.73	38.65	77.29
Air	Inpinger Train, BSOB	9.22	46.08	92.15
Air	Infrared Analyzer, Miran 1B	456.00	2280.00	4560.00
Air	Manometer, Micro Digital	10.45	52.25	104.50
Air	Meteorgraph	13.30	66.50	133.00
Air	Microscope, BHT Laboratory	39.35	196.74	393.47
Air	Orsat Gas Apparatus, Type D	16.70	83.51	167.01
Air	Pitot Tube, 36″ × 5/16″ diameter	0.76	3.81	7.62
Air	Pump, BSOB High Volume	4.54	22.71	45.41
Air	Pump, Calibration Kit, Gilian, HFS 113 AUC/5 Pumps	91.20	456.00	912.00
Air	Pump, Calibration Kit, Gilian, HFS 113 AUP/5 Pumps	95.00	475.00	950.00
Air	Pump, Gast High Volume	7.60	38.00	76.00
Air	Pump, Gast Low Volume	6.65	33.25	66.50
Air	Pump, Ind. Hy. Kit, Gilian, HFS 113 AUC/5 Pumps	84.55	422.75	845.50

(continued)

Equipment Category	Description	Daily	Weekly	Monthly
Air	Recorder, Linear	$ 3.80	$ 19.00	$ 38.00
Air	Rotameter (10-250 cc)	1.80	8.98	17.96
Air	Rotameter (200-4500 cc)	1.80	8.98	17.96
Air	Rotameter, Low Flow (Balt. Air Type)	3.95	19.76	39.52
Air	Sampler Kit, High Volume, Biss	35.15	175.75	351.50
Air	Sampler, Dioxin	20.90	104.50	209.00
Air	Sampler High Volume, Particulate	24.61	123.03	246.05
Air	Sampler, High Volume, Pesticide (PUF)	38.00	190.00	380.00
Air	Sampler, Monitaire, Personnel	10.17	50.83	101.65
Air	Sampler, 5-20 LPM Air System	17.01	85.03	170.05
Ancillary	AC Generator, Honda Portable 3500 W	14.91	74.57	149.15
Ancillary	AC Generator, Sears Portable 1100 W	5.23	26.13	52.25
Ancillary	Battery Charger, 12 V	1.08	5.41	10.83
Ancillary	Battery Charger, 6 V/12 V	1.80	9.02	18.05
Ancillary	Battery, 12 V Heavy Duty Marine	1.63	8.17	16.34
Ancillary	Camera, 35 mm Richo	1.52	7.60	15.20
Ancillary	Drill Portable, 3/8 in.	0.95	4.75	9.50
Ancillary	Equipment Containers, Cartpak	0.48	2.38	4.75
Ancillary	Equipment Shelters, NPS, 10 gauge steel	9.50	47.50	95.00
Ancillary	Equipment Shelters, NPS, 14 gauge steel	7.60	38.00	76.00
Ancillary	Extension Cord, 100 ft., Grounded	.74	3.71	7.41
Ancillary	Extension Cord, 25 ft., Grounded	.32	1.62	3.23
Ancillary	Extension Cord, 50 ft., Grounded	.51	2.57	5.13
Ancillary	Extension Ladder, 20 ft.	1.90	9.50	19.00
Ancillary	Hand Level	1.33	6.65	13.30
Ancillary	Hand Sprayer, Decon	.57	2.85	5.70

Equipment Category	Description	Daily	Weekly	Monthly
Ancillary	Hand Truck, Light Duty	$ .32	$ 1.62	$ 3.23
Ancillary	Heater, Coleman Catalytic	.57	2.85	5.70
Ancillary	Heater, Portable Electric, 1500 W	.80	3.99	7.98
Ancillary	Optical Tape Measure	2.09	10.45	20.90
Ancillary	Pocket Rod	.24	1.19	2.38
Ancillary	Rain Gauge Recorder, Long-Term	19.95	99.75	199.50
Ancillary	Rain Gauge, Tipping Bucket with Recorder	22.80	114.00	228.00
Ancillary	Sabersaw, Portable	.86	4.28	8.55
Ancillary	Scale, Ohaus Model C151, Digital, Battery and AC Adapter	5.45	27.27	54.53
Ancillary	Stereoscope, 2-Power/4-Power Abrams Model CB-1	1.61	8.06	16.12
Ancillary	Survey Transit and Tripod, David White	36.82	184.10	368.20
Ancillary	Thermometer	10.45	52.25	104.50
Ancillary	Tool Kit	7.60	38.00	76.00
Ancillary	Thermometer, Micro Digital	10.45	52.25	104.50
Ancillary	Tool Kit	7.60	38.00	76.00
Ancillary	Water Bath, Polypropylene w/ss Lid	10.83	54.15	108.30
Ancillary	Weight Set, Calibrated, Type S, 1 mg-100 mg	5.40	26.98	53.96
Ancillary	Weight, 10 gram, Class "S"	.37	1.85	3.71
Ancillary	Weight, 10 gram, IOLM	.01	.07	.14
Ancillary	Weight, 50 gram, IOLM	.02	.12	.25
Communication	Radio System, G.E., with Charger, etc.	22.80	114.00	228.00
Communication	Radio System, Maxon, with Charger, etc.	21.32	106.58	213.15
Communication	Radio System, Motorola, with Charger, etc.	12.44	62.22	124.45
Communication	Radio, Weather Band	.57	2.85	5.69
Groundwater	Bailer, 1.05″ × 3′, Fluorocarbon	4.45	22.23	44.46

(continued)

Equipment Category	Description	Daily	Weekly	Monthly
Groundwater	Bailer, 1.66″ × 3′, Fluorocarbon	$ 4.71	$ 23.56	$ 47.12
Groundwater	Bailer, 3.5″ × 3′, Fluorocarbon	6.57	32.87	65.74
Groundwater	Bailer, Teflon; 3″ × 3′ Double Check Valve	5.93	29.64	59.28
Groundwater	Bladder Pump, Teflon	8.84	44.18	88.35
Groundwater	Control Box, Bladder Pump	10.85	54.24	108.49
Groundwater	Field Printer, Model SE1004B	10.45	52.25	104.50
Groundwater	Filter, Pressure, 2 L Stainless Steel	23.81	119.04	238.07
Groundwater	Hermit Data Logger	47.56	237.79	475.57
Groundwater	Portable ISCO Pump	11.69	58.43	116.85
Groundwater	Pressure Transmitter, Model PTX 160 D	20.43	102.13	204.25
Groundwater	Pump, Centrifugal, 2.2, 1.5″	8.17	40.85	81.70
Groundwater	Pump, Gast Vacuum and Filter Apparatus	1.14	5.70	11.40
Groundwater	Pump, Hand Vacuum and Acc.	.95	4.75	9.50
Groundwater	Pump, Peristaltic	12.25	61.27	122.55
Groundwater	Pump, Submersible, 12 V	72.11	360.53	721.05
Groundwater	Pump, Submersible, Jacuzzi, 4″	7.13	35.63	71.25
Groundwater	Stainless Steel Tape with Weight	4.38	21.91	43.83
Groundwater	Tape Measure, Steel, 100 ft.	2.58	12.92	25.84
Groundwater	Tape Measure, Steel, 200 ft.	4.66	23.28	46.55
Safety	Air Mask, MSA Units, Ultralite, w/Airline Hookup	24.70	123.50	247.00
Safety	Air Mask, MSA Units, Ultralite, w/o Airline Hookup	23.94	119.70	239.40
Safety	Air Mask, Survivair Units	19.00	95.00	190.00
Safety	Airline Respirator with 5-Minute Tank	14.25	71.25	142.50
Safety	Belt, Orange Dacron, Single D Ring	1.28	6.42	12.83
Safety	Detector Pump Kit	4.75	23.75	47.50
Safety	Explosimeter Combustible Gas Ind.	5.32	26.60	53.20

Equipment Category	Description	Daily	Weekly	Monthly
Safety	Explosimeter, MSA Model 2A	$ 5.32	$ 26.60	$ 53.20
Safety	FID Detector Recorder, Foxboro OVA-128	11.40	57.00	114.00
Safety	FID Detector, Foxbore OVA-128	98.80	494.00	988.00
Safety	FID Detector, Foxboro OVA-128 with GC	115.90	579.50	1,159.00
Safety	Fire Extinguisher, 5 lb. Dry Chemical	.51	2.56	5.13
Safety	First Aid Kit, Light Industrial	1.28	6.39	12.78
Safety	First Aid Kit, Refill	.32	1.61	3.22
Safety	HCN Detector, 10 PPM with Alarm	18.43	92.15	184.30
Safety	MK-2 Audio Dosimeter, DuPont	24.32	121.60	243.20
Safety	Powered Air Purifying Respirator	10.75	53.76	107.52
Safety	PID Detector Probe, 10.2 eV	38.00	190.00	380.00
Safety	PID Detector Probe, 11.7 eV	38.00	190.00	380.00
Safety	PID Detector Recorder, HNU	11.40	57.00	114.00
Safety	PID Detector, HNO Photoionizer, PI-101	71.06	355.30	710.60
Safety	PID, Photovac Tip ''I''	61.75	308.75	617.50
Safety	PID, Photovac Tip ''II''	74.10	370.50	741.00
Safety	Quick Fix	9.40	47.02	94.05
Safety	SCBA Tank, MSA Ultralite, 30 min.	5.04	25.18	50.35
Safety	SCBA Tank, Survivair, 30 min.	3.80	19.00	38.00
Safety	Survey Meter, Model 3 G-M	5.80	28.98	57.95
Safety	Total Encapulating Suit, Vautex, SCBA	52.82	264.10	528.20
Soil/Sludge	Auger, Bucket Type, Stainless Steel	7.60	38.00	76.00
Soil/Sludge	Concrete Coring Drill	36.10	180.50	361.00
Soil/Sludge	Dredge, Ekman	4.09	20.43	40.85
Soil/Sludge	Sludge/Sediment Sampler	7.50	37.52	75.05

(continued)

Equipment Category	Description	Daily	Weekly	Monthly
Soil/Sludge	Soil Coring Kit	$ 5.28	$ 26.41	$ 52.82
Soil/Sludge	Split Spoon Sampler	17.10	85.50	171.00
Surface Water	Conductivity Meter, Temp Comp.	5.49	27.46	54.91
Surface Water	Dissolved Oxygen Meter YSI	15.11	75.53	151.05
Surface Water	Flowmeter, ISCO Model 1870	46.55	232.75	465.50
Surface Water	Kemmerer Bottle, Stainless Steel, 3.2 liter	4.37	21.87	43.74
Surface Water	Oxygen Meter, YSI Model 54	12.25	61.27	122.55
Surface Water	Oxygen Meter, YSI Model 57	12.25	61.27	122.55
Surface Water	pH Meter, Digi-Sense	5.23	26.13	52.25
Surface Water	Proms, Primary Flow Modules	2.85	14.25	28.50
Surface Water	S-C-T Meter, YSI Model 33 with Probe	9.22	46.08	92.15
Surface Water	Sampler, Composite, ISCO Model 1590	19.57	97.85	195.70
Surface Water	Sampler, Discrete, ISCO Model 1680	28.88	144.40	288.80
Surface Water	Turbidimeter, Hach, 2100A	19.09	95.47	190.95
Surface Water	Velocity Meter, Marsh McBirney 250 VMFM, Model 210	47.50	237.50	475.00
Surface Water	Velocity Meter, Pygmy Rod, w/Digital Readout	40.85	204.25	408.50
Surface Water	Velocity Meter, Ultrasonic, Unimeter	5.61	28.03	56.05

TYPICAL PRICES FOR CONSUMABLE SUPPLIES

Item	Cost	Unit	Count	Cost/ Unit
Activated Charcoal Tubes	$ 27.25	Case	36	$ .76
Anti-Fog Solution, Fogpruf, 4 oz.	2.35	Each	1	2.35
Bags, 4 ml Poly, 22-1/2″ × 20″ × 48″	45.01	Case	75	.60
Bags, 6 ml Poly, 12″ × 16″	60.20	Case	500	.12
Bags, 9″ × 12″ Zip Lip, 2 mil	57.96	Pack	200	.29
Bags, Poly, Cooler Liners	53.80	Case	100	.54
Bags, Ziplocks 5″ × 8″	25.83	Case	500	.05
Blades, Surgical, Size 22, Length 55 mm	52.00	Box	75	.69
Bootcovers, Yellow PVC	120.00	Case	60	2.00
Boots, Kaysam Bulldog, Size 12	10.37	Each	1	10.37
Bucket, Stainless Steel, 23.25 qt.	77.20	Each	1	77.20
Cap Liner	5.00	Each	1	5.00
Cap, White V-Gard Slotted	6.05	Each	1	6.05
Capillary Scoring Pencil	10.00	Each	1	10.00
Caps, 15 ml Teflon Lined	30.00	Case	144	.21
Caps, 20 mm Teflon Lined	100.00	Case	144	.69
Cartpak Containers with Lids, 15″ × 21″ × 27″	38.92			
Cartridge, Resp, Type H	36.80	Each	10	3.68
Chain of Custody Forms	428.00	Case	3000	.14
Charging Hose	159.50	Each	1	159.50
Clamp Holders	6.00	Each	1	6.00
Clamps, Three Prong, 3.5″	11.95	Each	1	11.95
Cleaner/Sanitizer	13.60	Case	12	1.13
Conductive Heel Strap, Statfree	22.00	Case	100	.22
Container, Amber Glass, 1000 ml	37.92	Case	12	3.16
Container, Amber Glass, 4000 ml	21.00	Case	4	5.25
Container, Amber Glass, 500 ml	27.60	Case	12	2.30
Container, Clear Glass, 250 ml Tall	34.32	Case	12	2.86
Container, Clear Glass, 500 ml Short	34.20	Case	12	2.85
Container, Clear Glass, 500 ml Tall	35.04	Case	12	2.92
Container, Clear Glass, 1000 ml	43.08	Case	12	3.59
Container, Clear Glass, 125 ml Tall	33.12	Case	12	2.76
Container, Clear Glass, 250 ml Short	33.60	Case	12	2.80
Container, *in-situ*, 500 ml Nalgene	20.90	Each	12	1.74
Container, Plastic Wide Mouth, 1000 ml	37.08	Each	12	3.09
Container, Plastic, 1000 ml	32.88	Case	12	2.74

(continued)

Item	Cost	Unit	Count	Cost/ Unit
Container, Vial, Teflon Septa, 40 ml	$ 93.60	Case	72	$ 1.30
Coolers, Igloo, 48 qt	23.96	Each	1	23.96
Coupling Assembly, Flexible	98.50	Each	1	98.50
Dispensers, 2 ml Single Volume	53.46	Case	6	8.91
Dispensers, 5 ml Single Volume	53.46	Case	6	8.91
Draeger Tube, Ammonia	30.00	Each	10	3.00
Draeger Tube, Carbon Monoxide	36.85	Each	10	3.69
Draeger Tube, Cyanide	36.00	Each	10	3.60
Draeger Tube, Hydrocarbon	32.00	Each	10	3.20
Draeger Tube, Hydrogen Sulfide	36.00	Each	10	3.60
Draeger Tube, Nitrogen Dioxide	28.00	Each	10	2.80
Draeger Tube, Perchloroethylene	32.00	Each	10	3.20
Draeger Tube, Sulfur Dioxide	39.15	Each	10	3.92
Draeger Tube, Vinyl Dioxide	32.00	Each	10	3.20
Faceshield Frame, Chemgard (less visor)	6.00	Each	1	6.00
Filter, Gelman 47 mm Type A/E	19.35	Box	100	.19
Filter, Glass High Purity (Hi-Vols)	27.50	Case	50	.55
Filters, Hi-Vol Impacter	150.00	Case	100	1.50
Flow Valve, Brass	7.83	Each	1	7.83
Formaldehyde Draeger Tubes	46.30	Each	10	4.63
Glasses, Safety Clear	26.50	Each	10	2.65
Glasses, Safety Grey	33.50	Each	10	3.35
Gloves, Cotton Liners	4.50	Box	12	.38
Gloves, Foodhandlers (PCB Wipe Sampling)	18.20	Case	1000	.02
Gloves, Neoprene	6.25	Each	1	6.25
Gloves, Nitrile	14.62	Case	12	1.22
Gloves, PVC	10.76	Case	100	.11
Gloves, Viton	18.75	Each	1	18.75
Hexane, Pesticide Grade, 4-1 gallon jugs	146.15	Case	4	36.54
Hydrocarbon Draeger Tubes	39.15	Each	10	3.92
Indicating Drierite, 10-20 Mesh	49.68	12 lb		
Indicating Drierite, 8 Mesh	20.80	lbs	5	4.16
Isopropanol, 4-1 Gallon Jugs	62.74	Case	4	15.69
Knife Handle #4 Stainless Steel	8.00	Each	1	8.00
Labels, Chain of Custody Seals	85.00	Roll	100	.09
Labels, Versar 3-Part	632.00	Case	10,000	.06
Labels–"Corrosive"	21.00	Roll	500	.04
Labels–"Orm-E Hazardous Substance"	72.00	Roll	1000	.07
Labels–"Oxidizer"	21.00	Roll	500	.04

Item	Cost	Unit	Count	Cost/ Unit
Lanyard, 6 foot with Dual Snaps	$ 34.95	Each	1	$ 34.95
Methanol, 4-1 Gallon Jugs	54.72	Case	4	13.68
Monogoggle with Vents	4.57	Each	1	4.57
Notebook, Write in the Rain	100.00	Each	10	10.00
Oxygen Membrane/KCL Kit	9.00	Each	1	9.00
Oxygen/Temperature Probe	125.60	Each	1	125.60
Paint Cans, 2 Gallon with Lids	114.00	Case	30	3.80
Paper, Strip, for Micromonitor SP-310	30.00	Box	6	5.00
pH Buffer – 10.00, Fisher, 500 ml	43.25	Each	6	7.21
pH Buffer – 4.0, Fisher, 500 ml	43.25	Each	6	7.21
pH Buffer – 7.0, Fisher, 500 ml	43.25	Each	6	7.21
pH Paper (1-12)	3.20	Each	1	3.20
Rain Jacket with Attached Hood and Bibs	20.46	Each	1	20.46
Respirator Filter, Type H, for Paper	36.80	Box	10	3.68
Respirator Fit Test Ampules	4.50	Each	10	.45
Respirator Spectacle Kit	26.10	Each	1	26.10
Respirator, Ultralite, Large	126.00	Each	1	126.00
Tape, 2″ Clear PVC	88.56	Case	36	2.46
Tape, 2″ Duct	100.32	Case	24	4.18
Tape, 3/4″ Strapping	72.00	Case	48	1.50
Thermometer, Mercury	17.25	Each	1	17.25
Towels, Kaydry	42.56	Case	18	2.36
Tubing Cutter, Imp	13.00	Each	1	13.00
Tubing, 1/8″ O.D. × 2.1 mm I.D. SS, 50′ Coil	61.00	Foot	50	1.22
Tyvex Suits, P.E. Coated	114.35	Case	25	4.57
Tyvex Suits, Saranex	372.50	Case	25	14.90
Visor, 6 inch Clear Polycarbonate (Topgards)	5.15	Each	1	5.15
Wash Bottle, 500 ml Teflon	51.27	Each	1	51.27
Water, Distilled, 4-1 Gallon Jugs	28.00	Case	4	7.00
Water, HPLC, 4-1 Gallon Bottles	65.00	Case	4	16.25

APPENDIX B

General Site Remediation Costs

THE tables and figures in this Appendix are used by permission of the American Petroleum Institute. Tables B.1-B.3 and Figures B.1-B.20 are reproduced from API Publication Number 4422, February 1986, *Cost Models for Selected Technologies for Removal of Gasoline Components from Groundwater*. Figures B.21-B.23 are reproduced from API Publication Number 4471, June 1988, *Treatment System for the Reduction of Aromatic Hydrocarbons and Ether Concentrations in Groundwater*.

TABLE B.1. Initial Assumed Concentrations of Soluble Hydrocarbons in the Contaminated Zone.

Benzene	24.73 mg/l
Toluene	14.08 mg/l
Xylene, total	9.38 mg/l
Ethylbenzene	1.81 mg/l
MTBE or TBA	10.00 mg/l

TABLE B.2. Random Walk Model Coefficients Used in the Development of the Cost Model.

	Case 1	Case 2	Case 3	Case 4
Transmissivity (gpd/ft)	1000	100	10	1
Hydraulic conductivity (gpd/ft^2)	100	10	1	0.1
Porosity	0.4	0.4	0.4	0.4
Storage coefficient	0.25	0.15	0.08	0.03
Retardation coefficients				
Benzene	4.4	4.4	4.4	4.4
Toluene	13.3	13.3	13.3	13.3
Ethylbenzene	36.3	36.3	36.3	36.3
Xylene	40.0	40.0	40.0	40.0
MTBE/TBA	1.0	1.0	1.0	1.0
Recovery well pump rate (gpm)	50	10	5	1

gpd = Gallons per day.
gpm = Gallons per minute.

TABLE B.3. Applicable Treatment Technologies as a Function of the Target Treatment Level.

Compound	Target Treatment Level			
	None	100 PPB	10 PPB	1 to 5 PPB
Benzene	Pump to sewer	Carbon adsorption or air stripping	Carbon adsorption	Air stripping/ carbon adsorption
Toluene	Pump to sewer	Carbon adsorption or air stripping	Carbon adsorption	Air stripping/ carbon adsorption
Xylene	Pump to sewer	Carbon adsorption or air stripping	Carbon adsorption	Air stripping/ carbon adsorption
Ethylbenzene	Pump to sewer	Carbon adsorption or air stripping	Carbon adsorption	Air stripping/ carbon adsorption
MTBE	Pump to sewer	Air stripping/ carbon adsorption possible	None proven	None proven
TBA	Pump to sewer	None proven *in-situ* biological possible	None proven *in-situ* biological possible	None proven *in-situ* biological possible

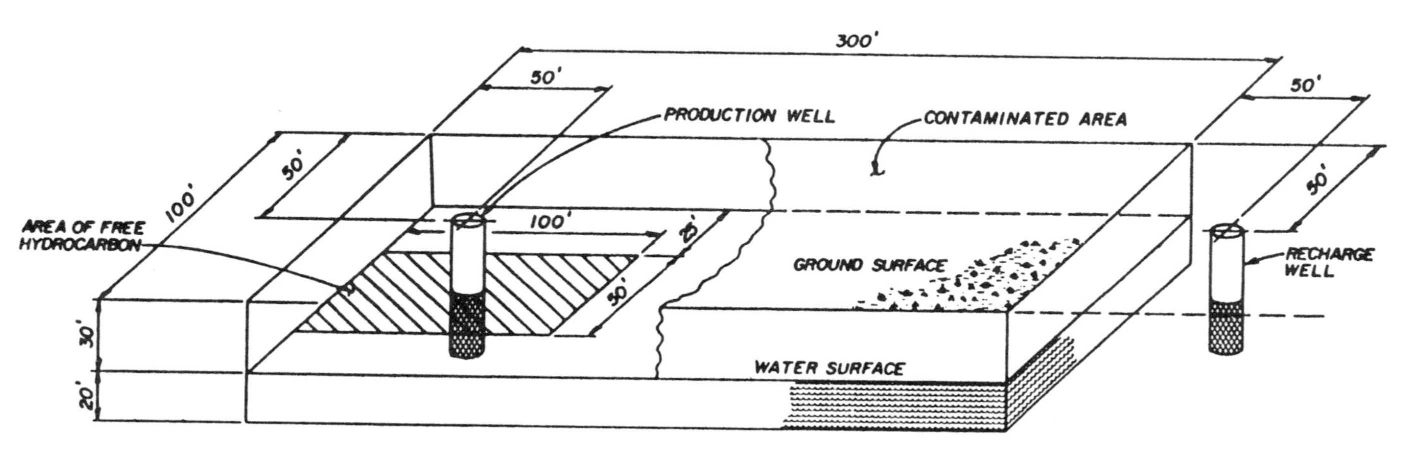

Figure B.1 Physical layout of contaminated area and well configuration.

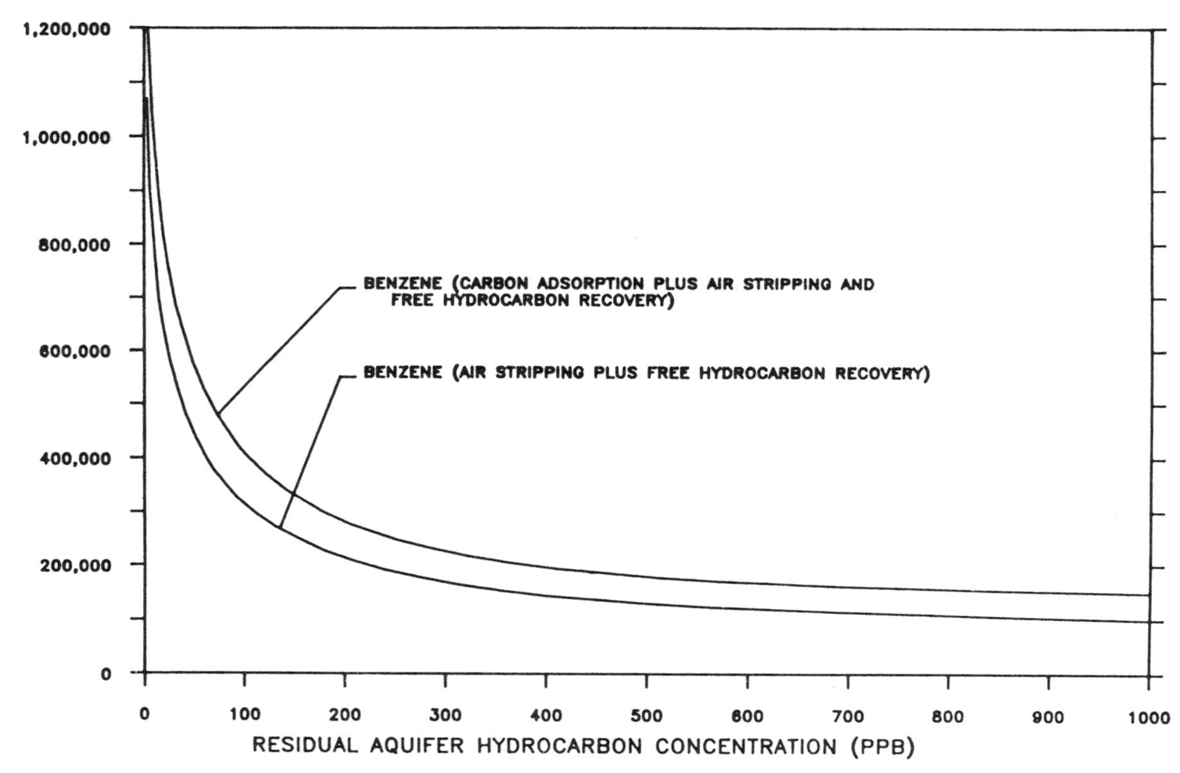

Figure B.2 Total aquifer restoration costs as a function of aquifer restoration level.

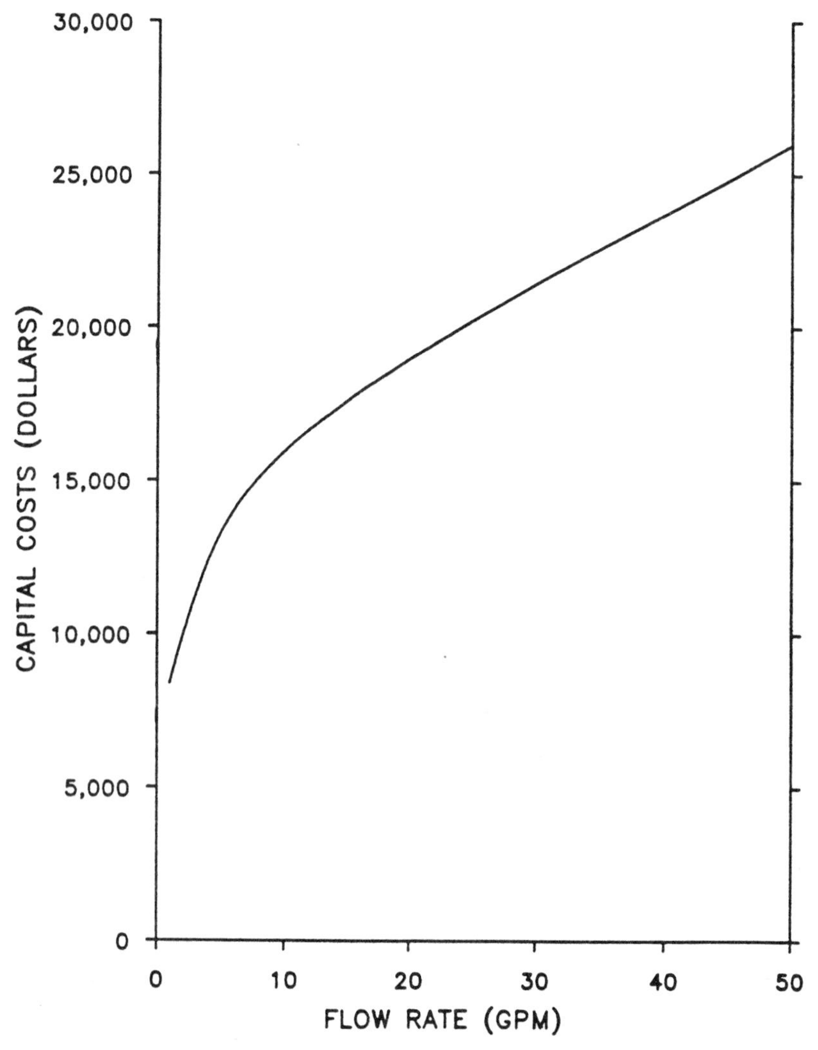

Figure B.3 Capital costs for pumping and phase separation.

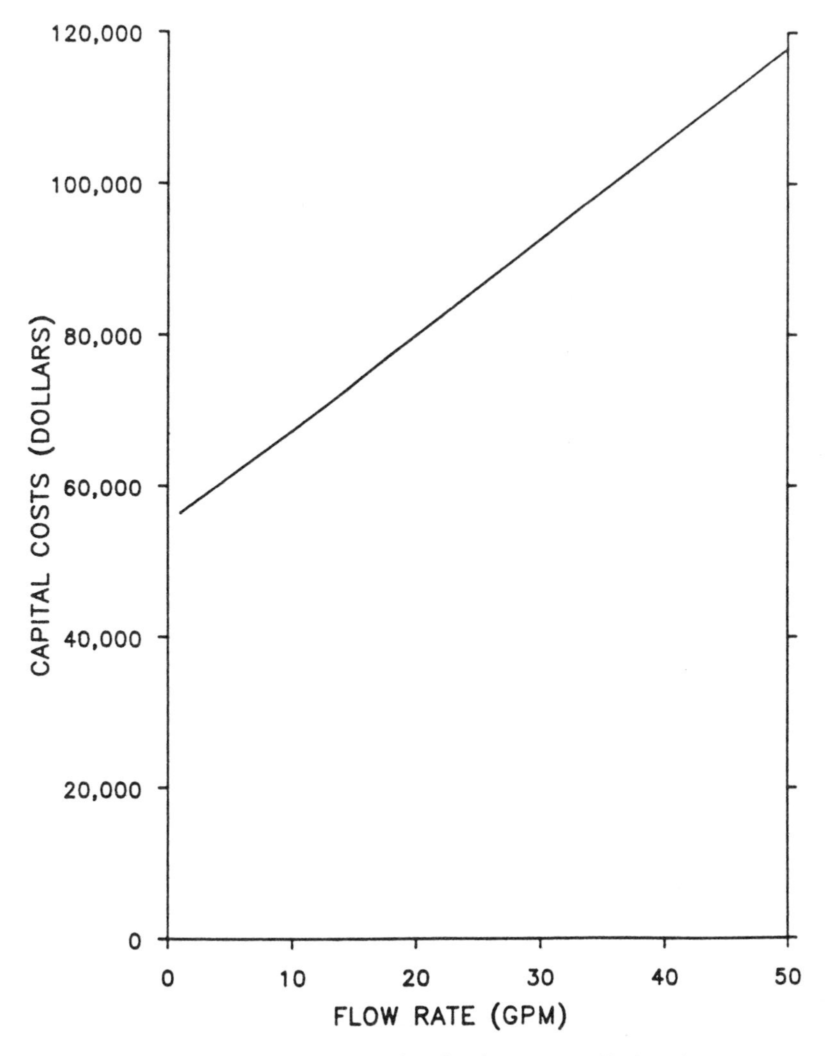

Figure B.4 Capital costs for carbon adsorption plus phase-separated hydrocarbon recovery.

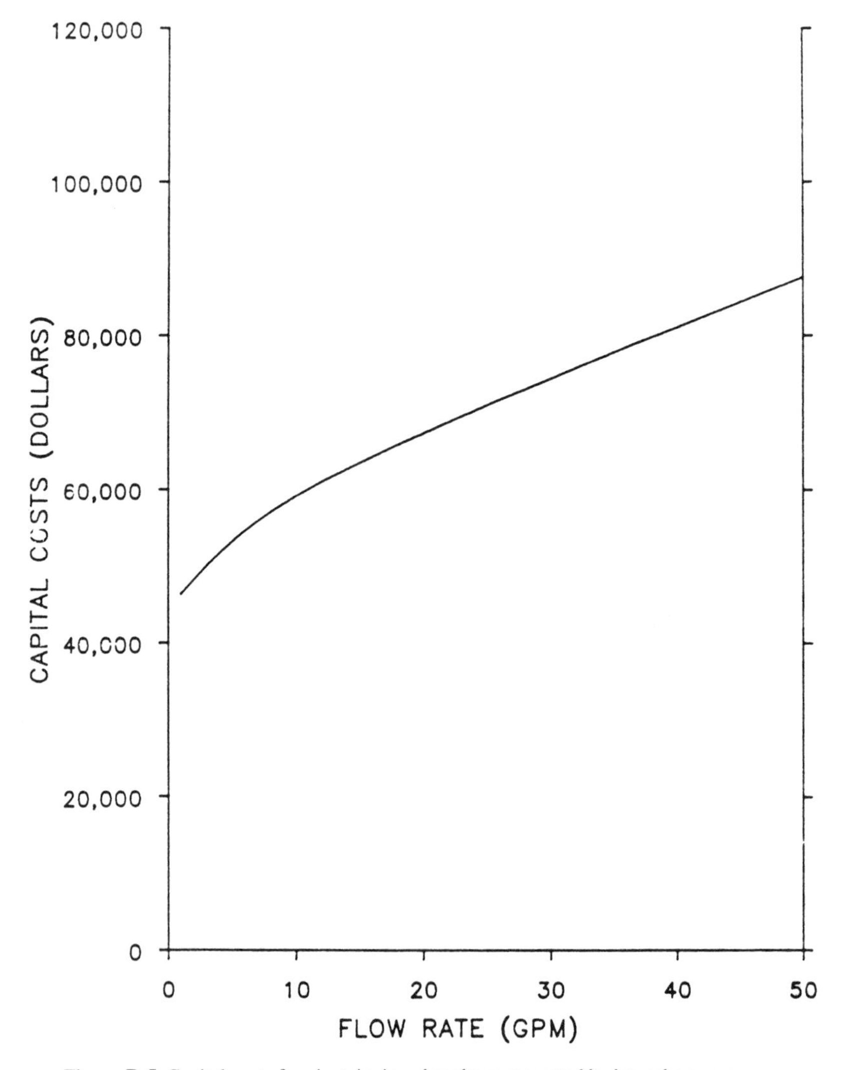

Figure B.5 Capital costs for air stripping plus phase-separated hydrocarbon recovery.

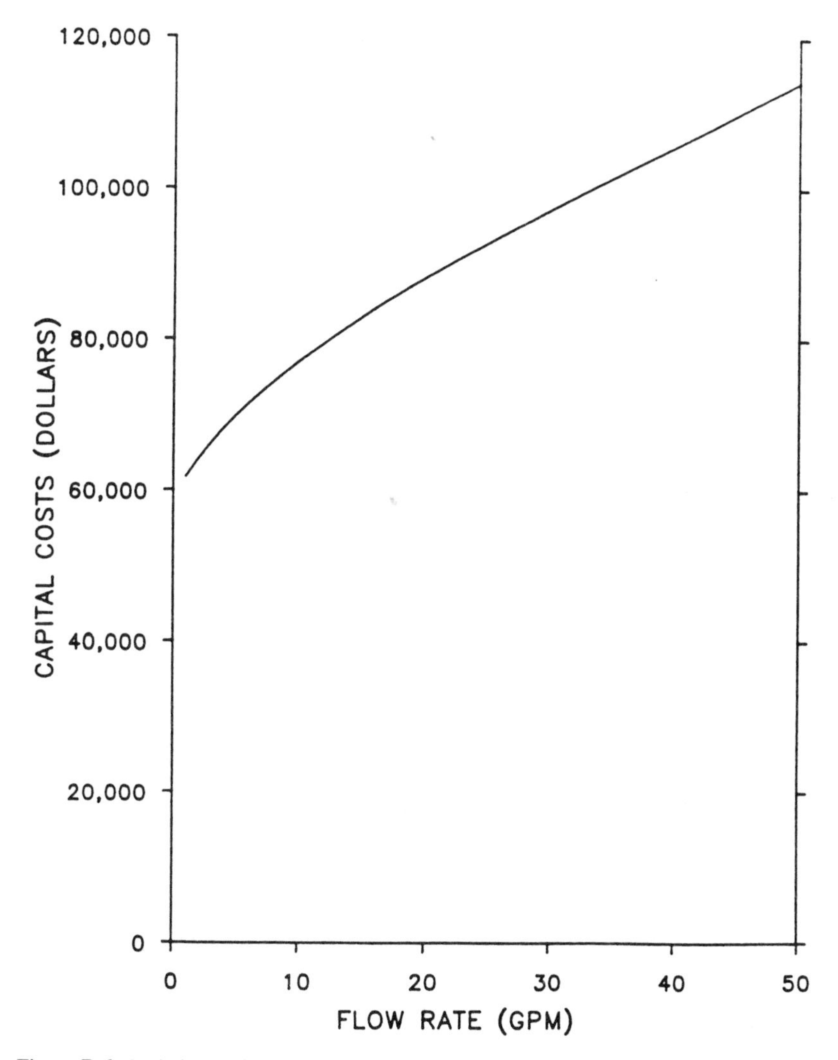

Figure B.6 Capital costs for air stripping/carbon adsorption plus phase-separated hydrocarbon recovery.

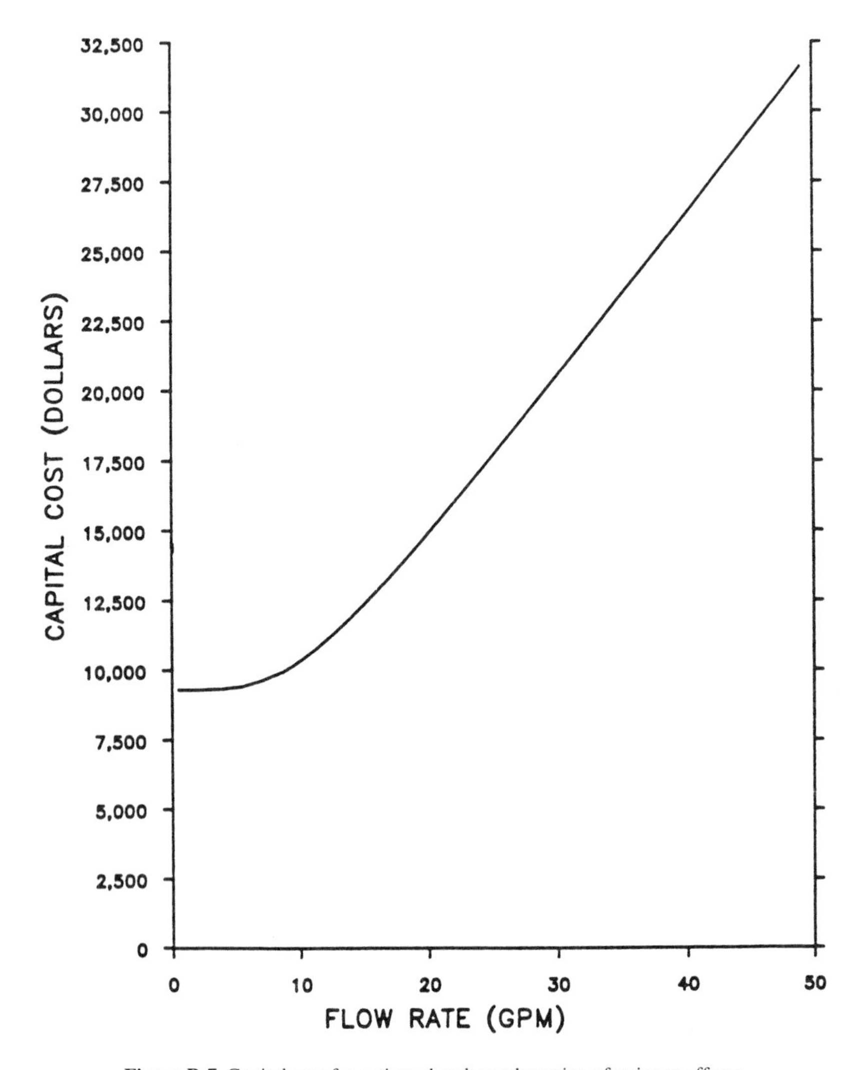

Figure B.7 Capital cost for activated carbon adsorption of stripper off-gas.

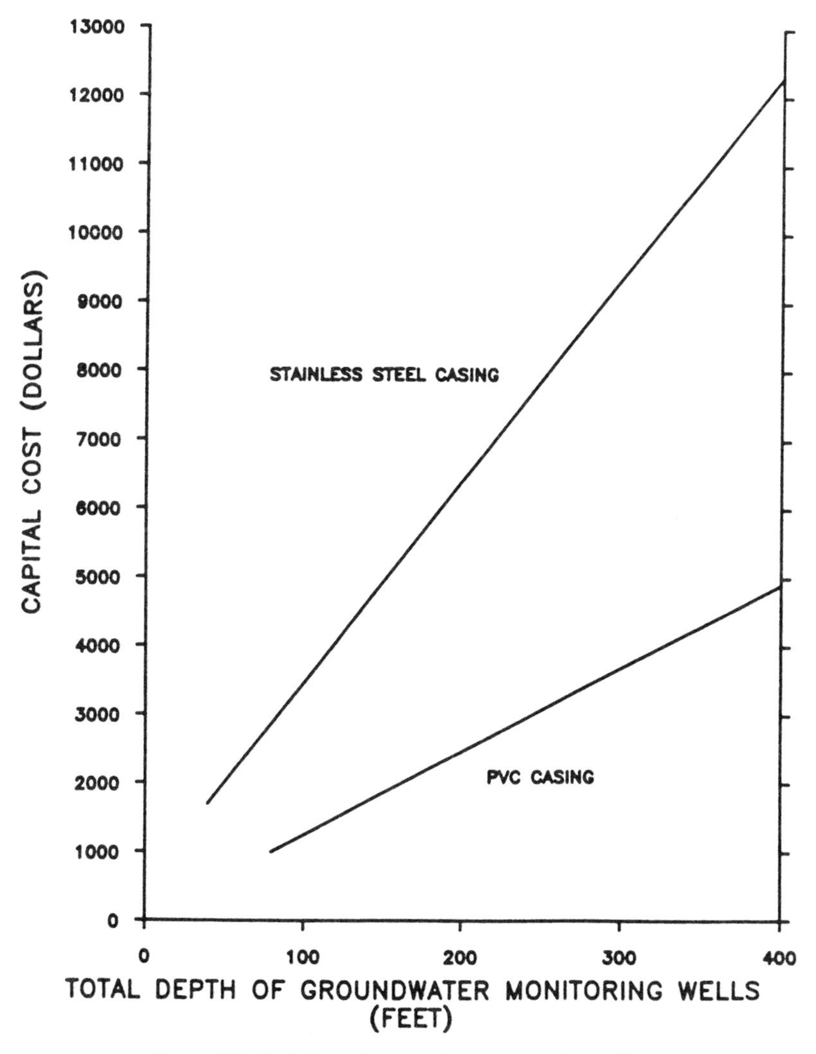

Figure B.8 Capital cost for groundwater monitoring wells.

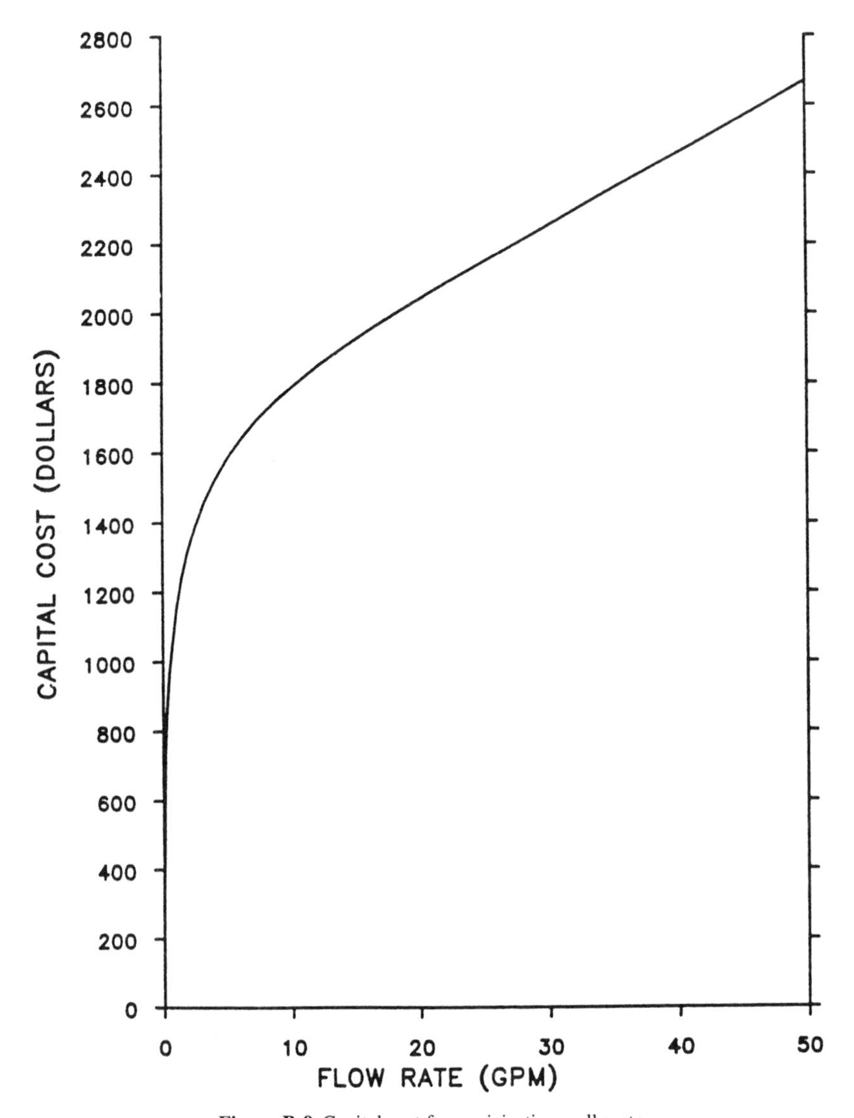

Figure B.9 Capital cost for a reinjection well system.

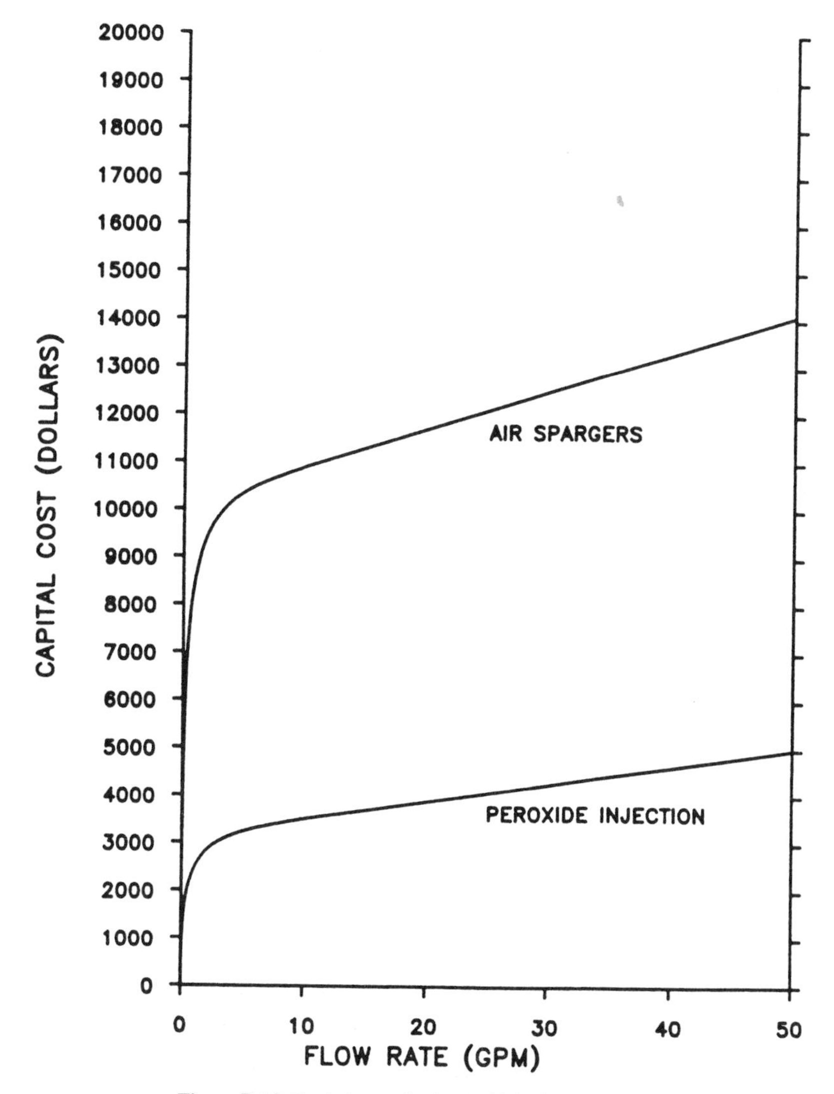

Figure B.10 Capital costs for *in-situ* biological treatment.

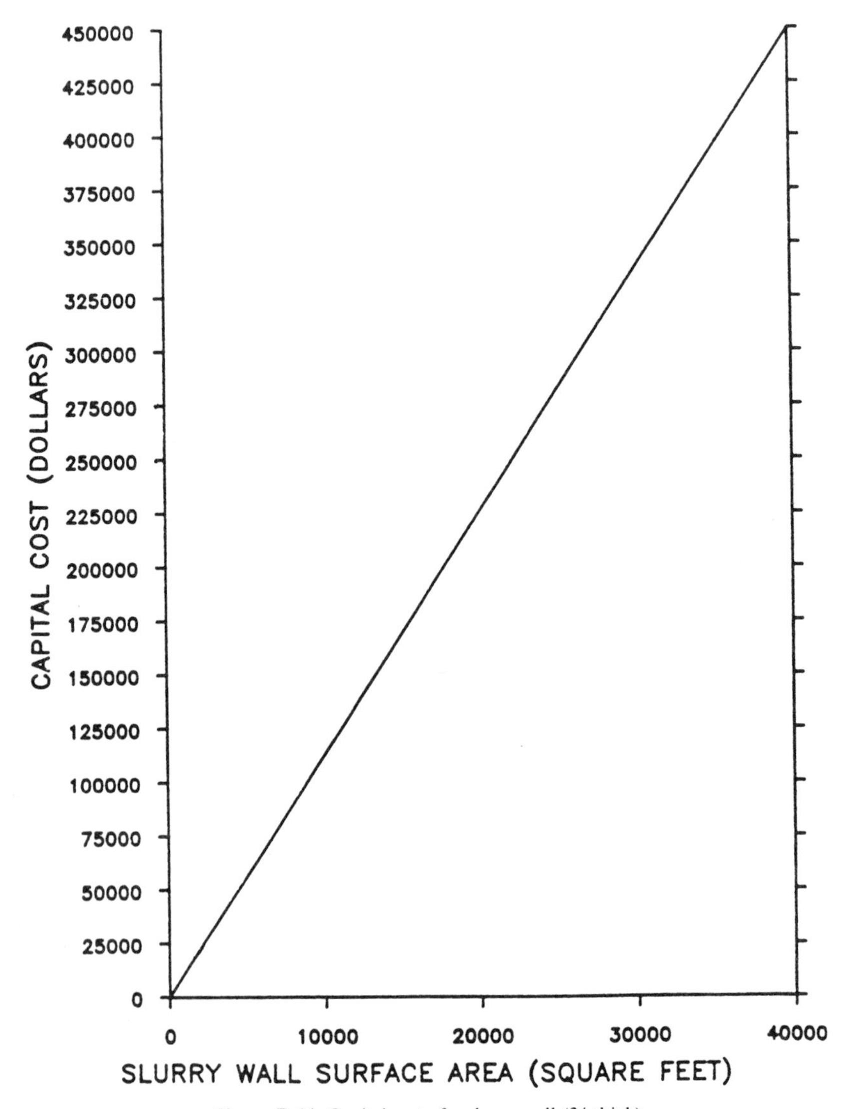

Figure B.11 Capital costs for slurry wall (3′ thick).

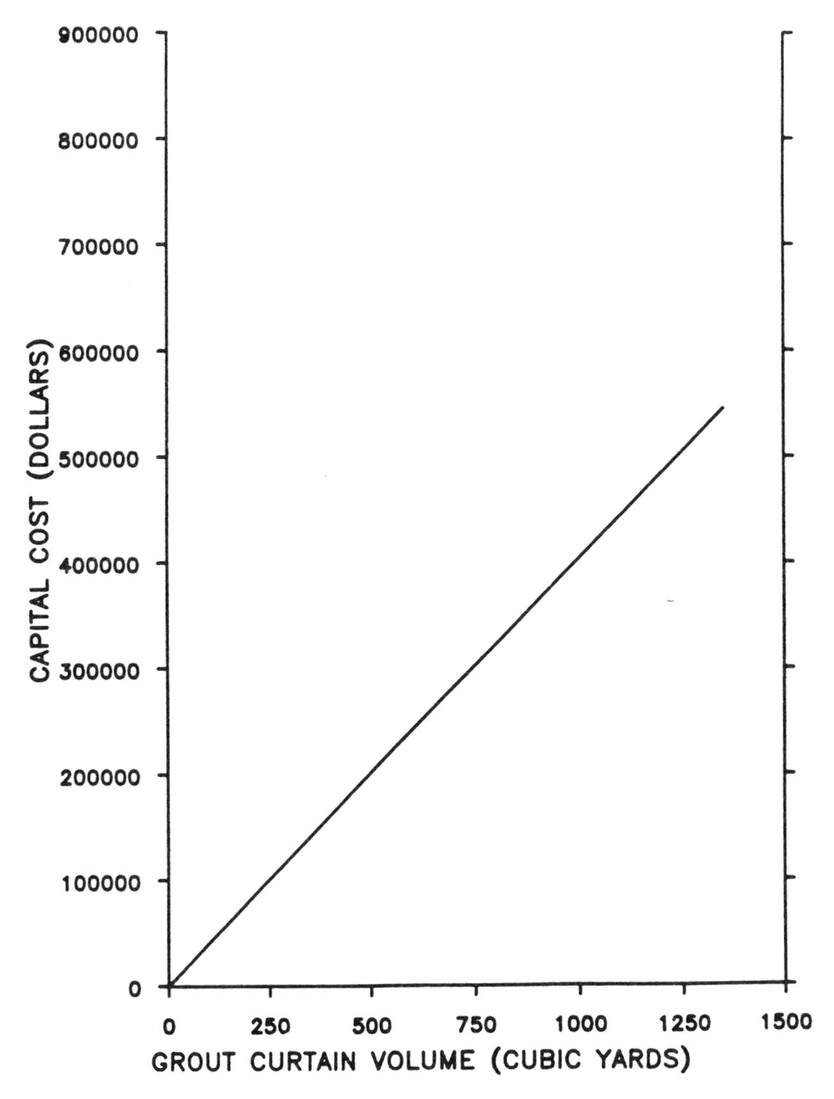

Figure B.12 Capital costs for grout curtain (3′ thick).

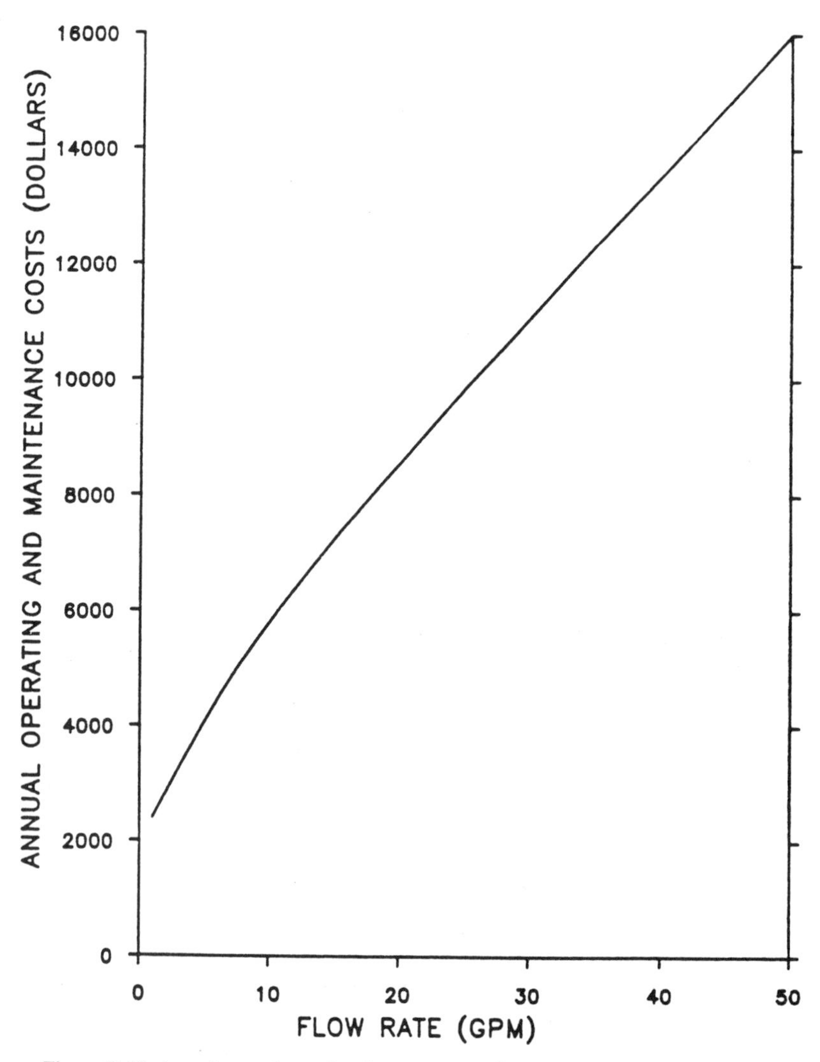

Figure B.13 Annual operating and maintenance costs for pumping and phase separation.

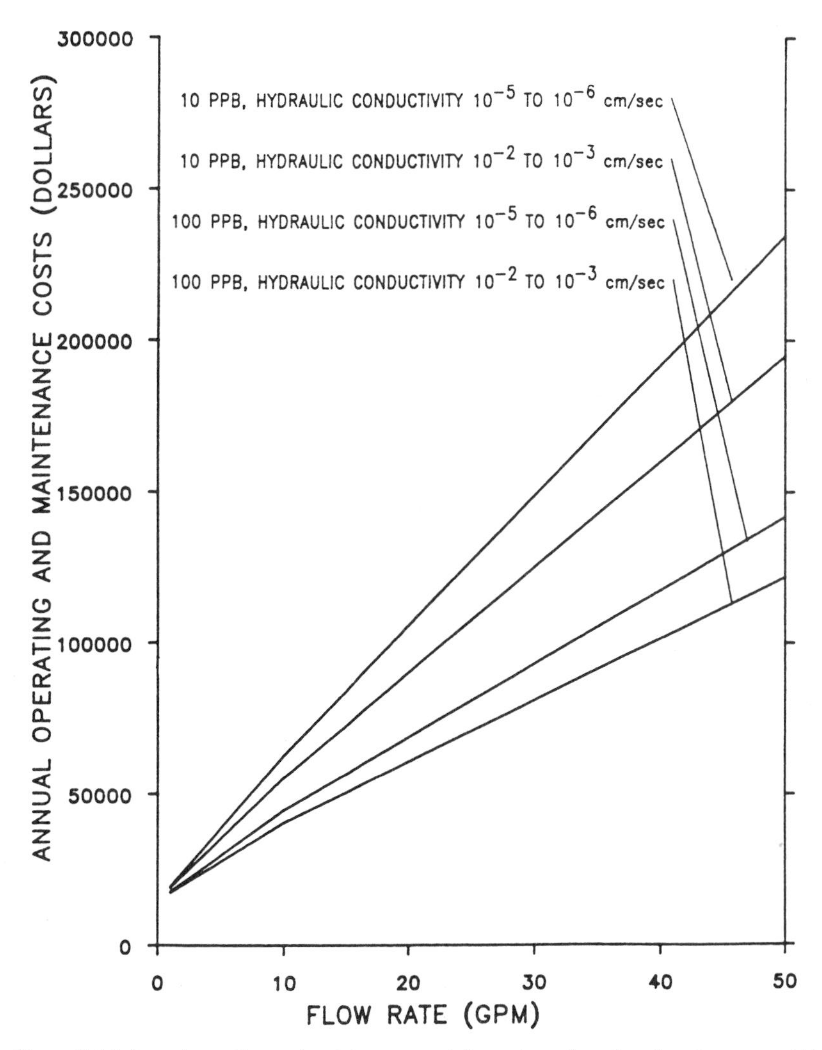

Figure B.14 Annual operating and maintenance costs for carbon adsorption plus phase-separated hydrocarbon recovery.

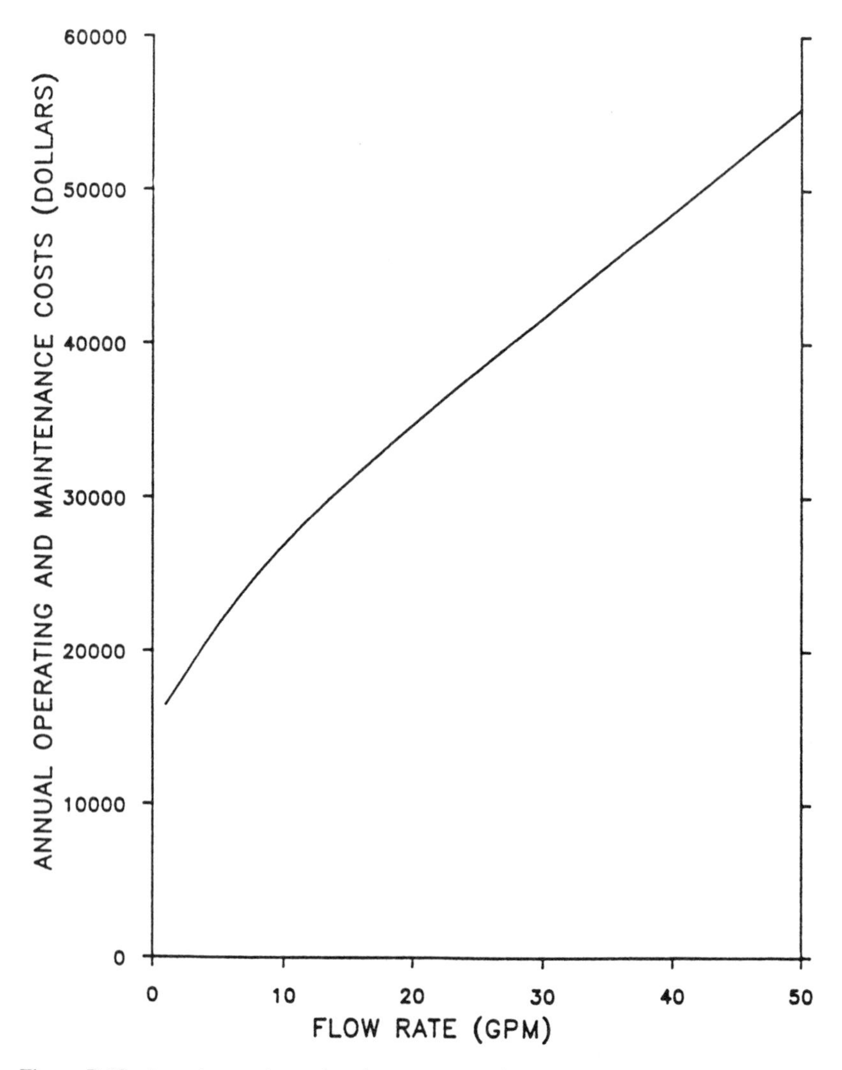

Figure B.15 Annual operating and maintenance costs for air stripping plus phase-separated hydrocarbon recovery.

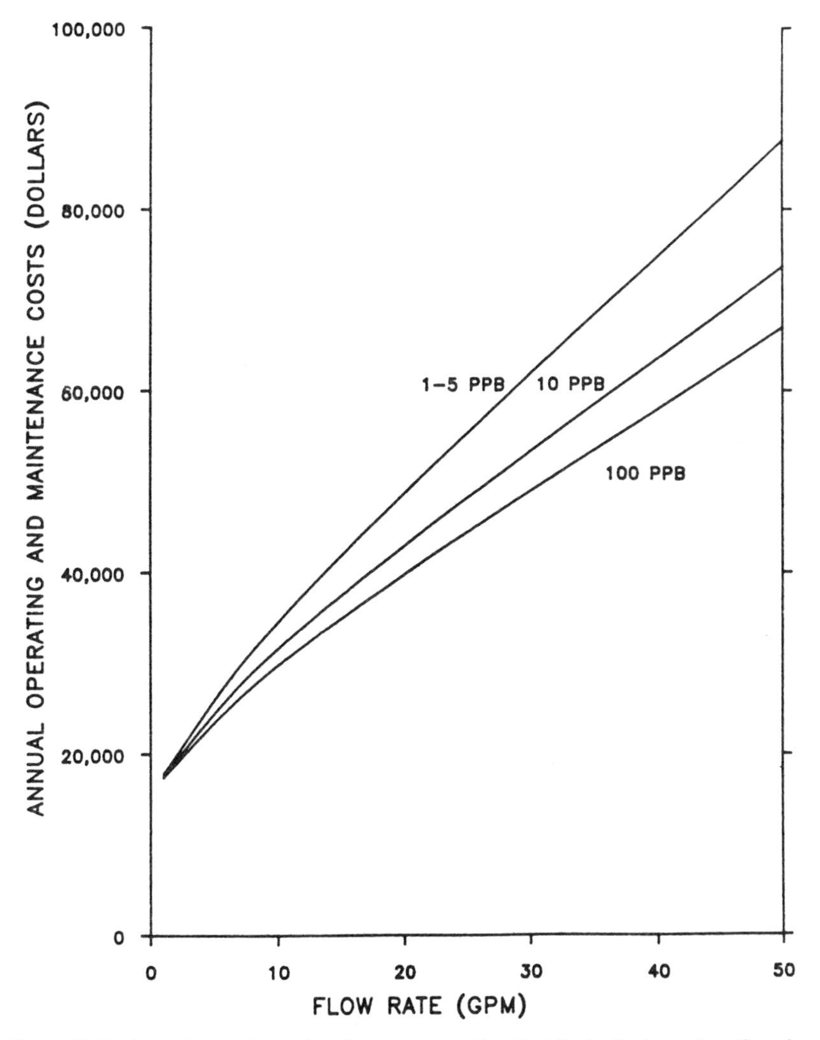

Figure B.16 Annual operating and maintenance costs for air stripping/carbon adsorption plus phase-separated hydrocarbon recovery.

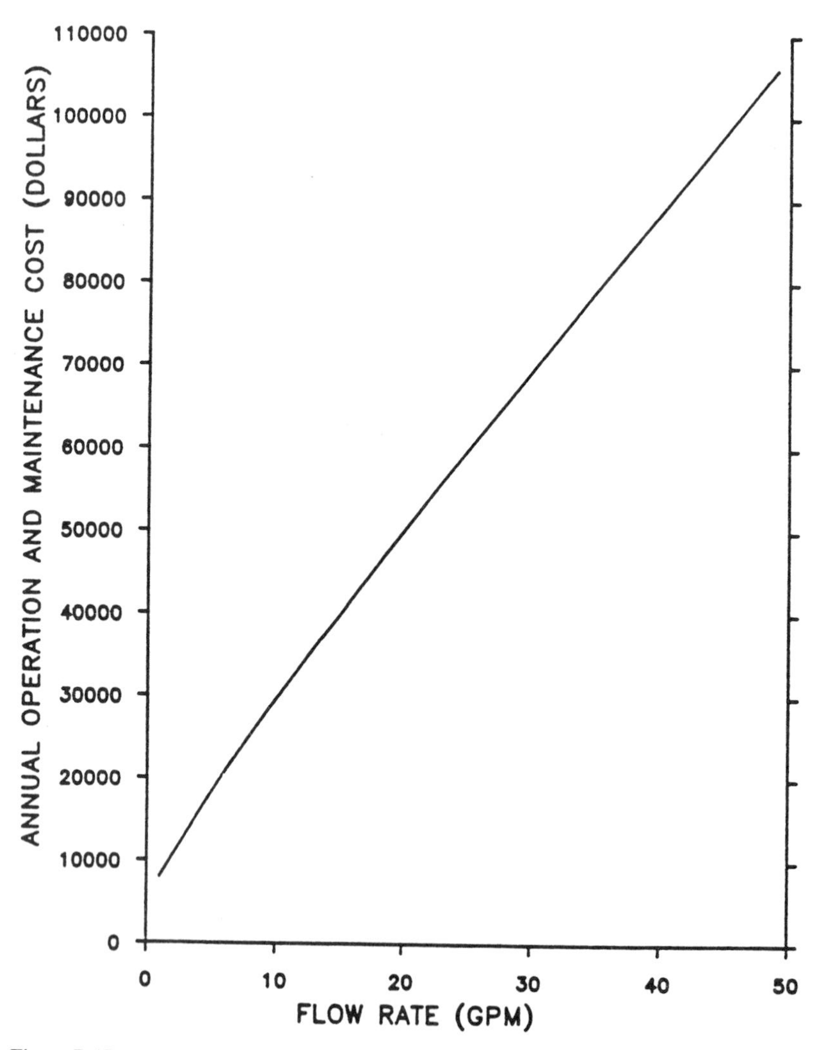

Figure B.17 Annual operation and maintenance costs for activated carbon adsorption of stripper off-gas.

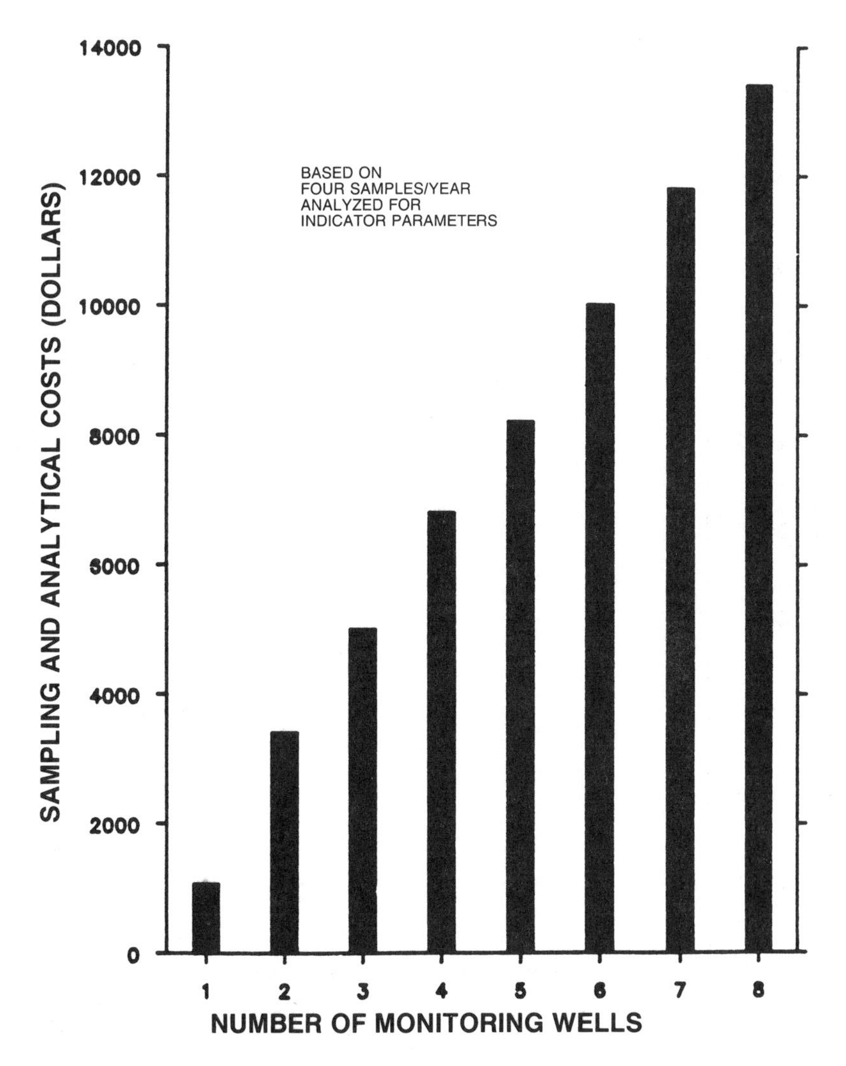

Figure B.18 Annual sampling and analytical costs for groundwater monitoring.

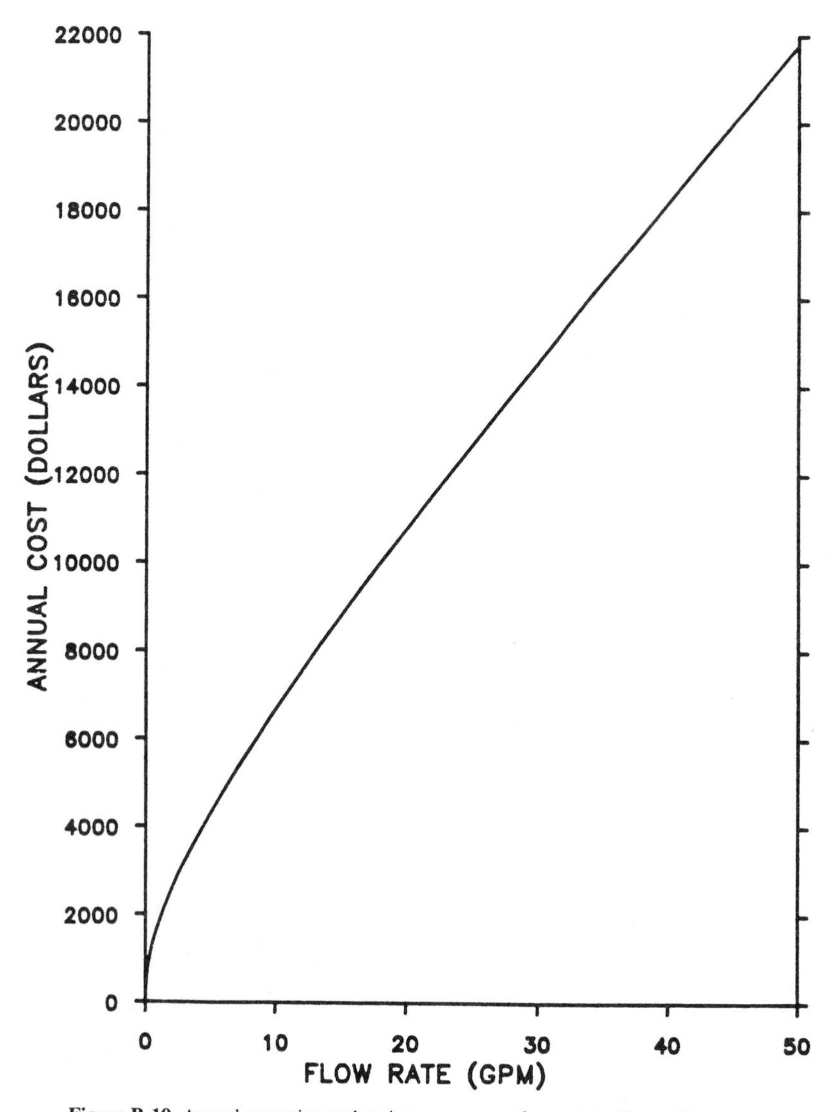

Figure B.19 Annual operation and maintenance costs for a reinjection well system.

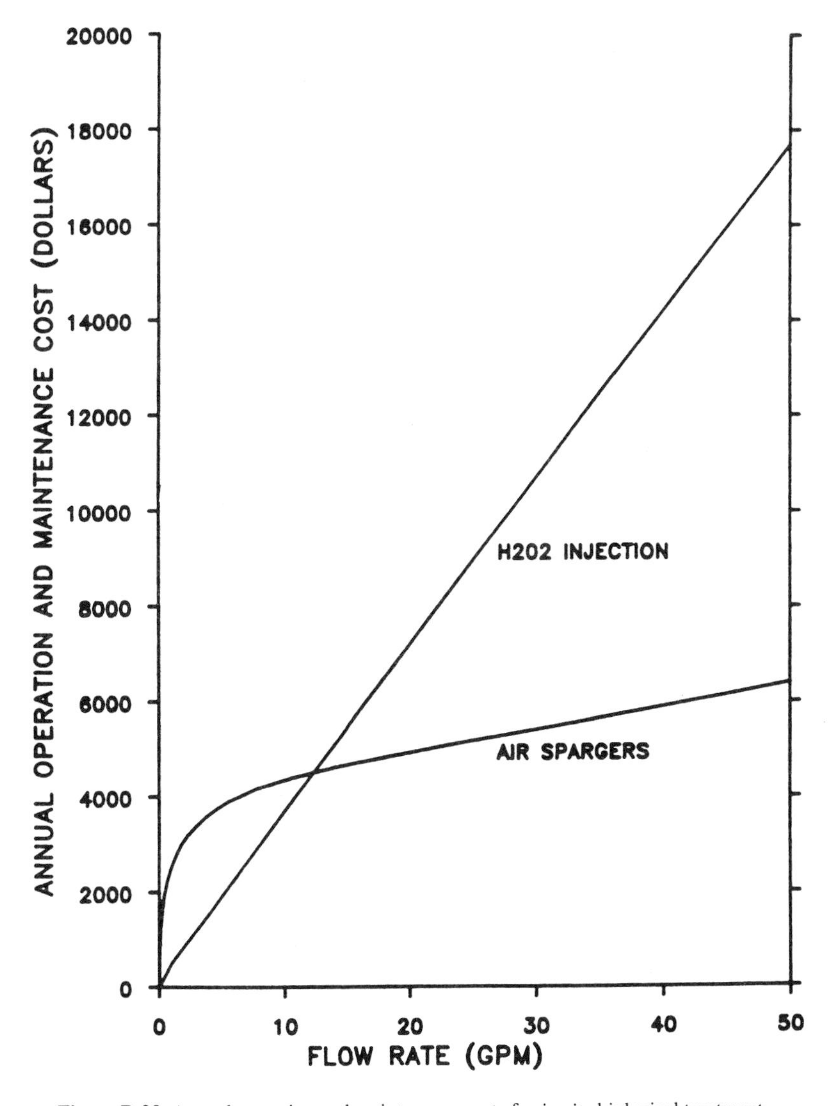

Figure B.20 Annual operation and maintenance costs for *in-situ* biological treatment.

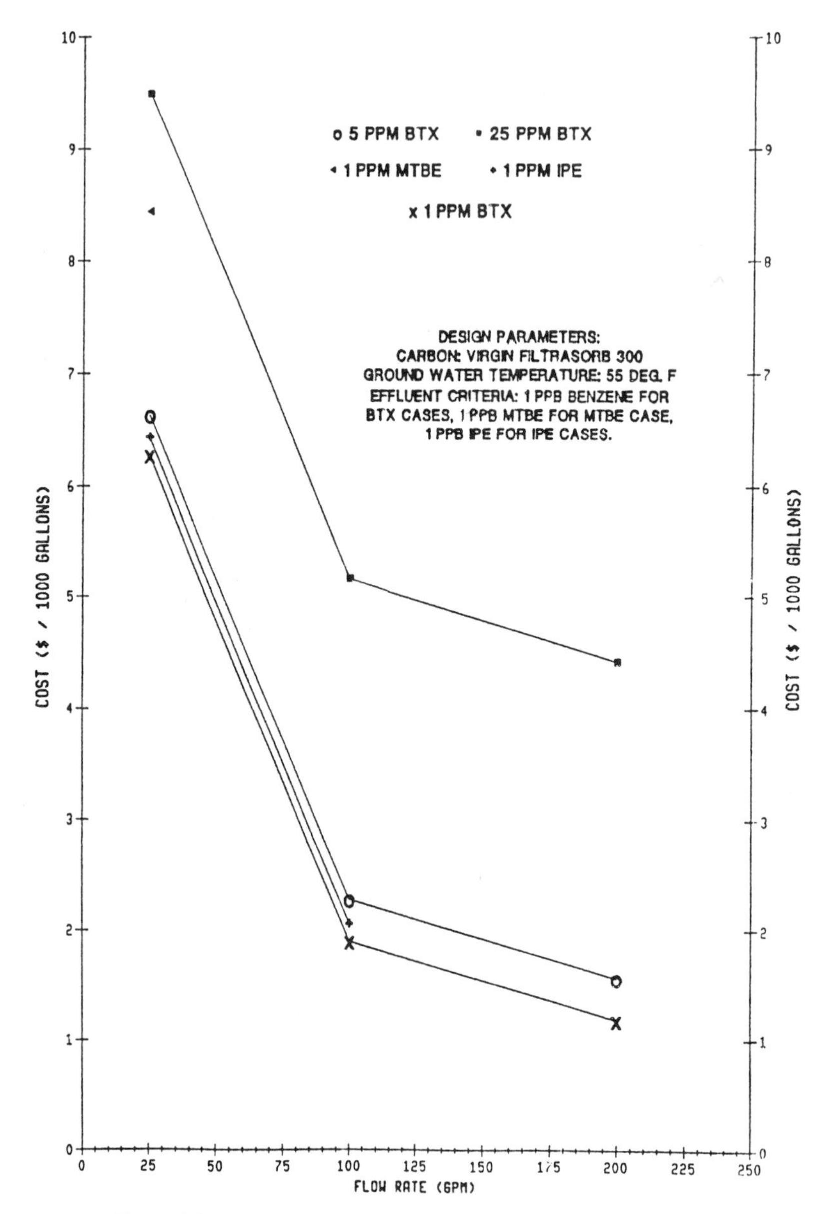

Figure B.21 Cost versus flow rate, carbon adsorption treatment systems.

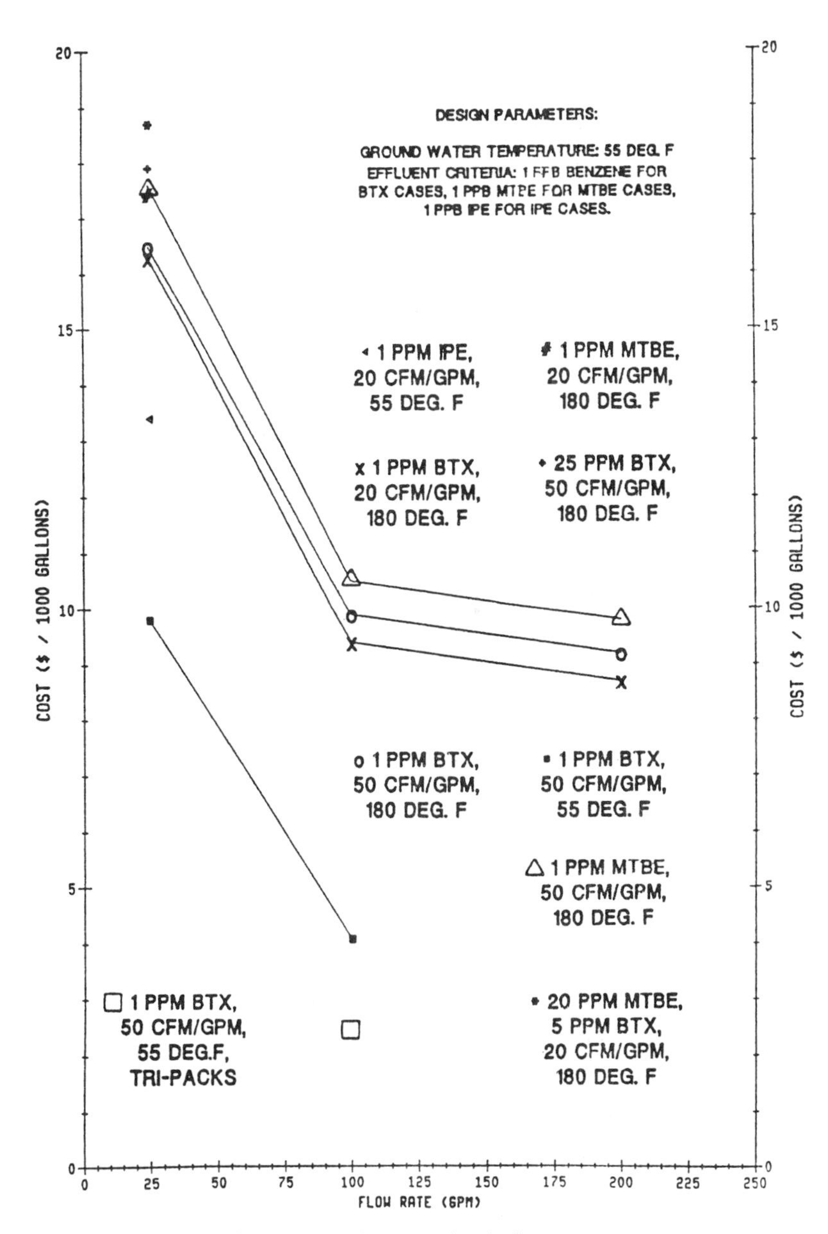

Figure B.22 Cost versus flow rate, air stripping treatment systems.

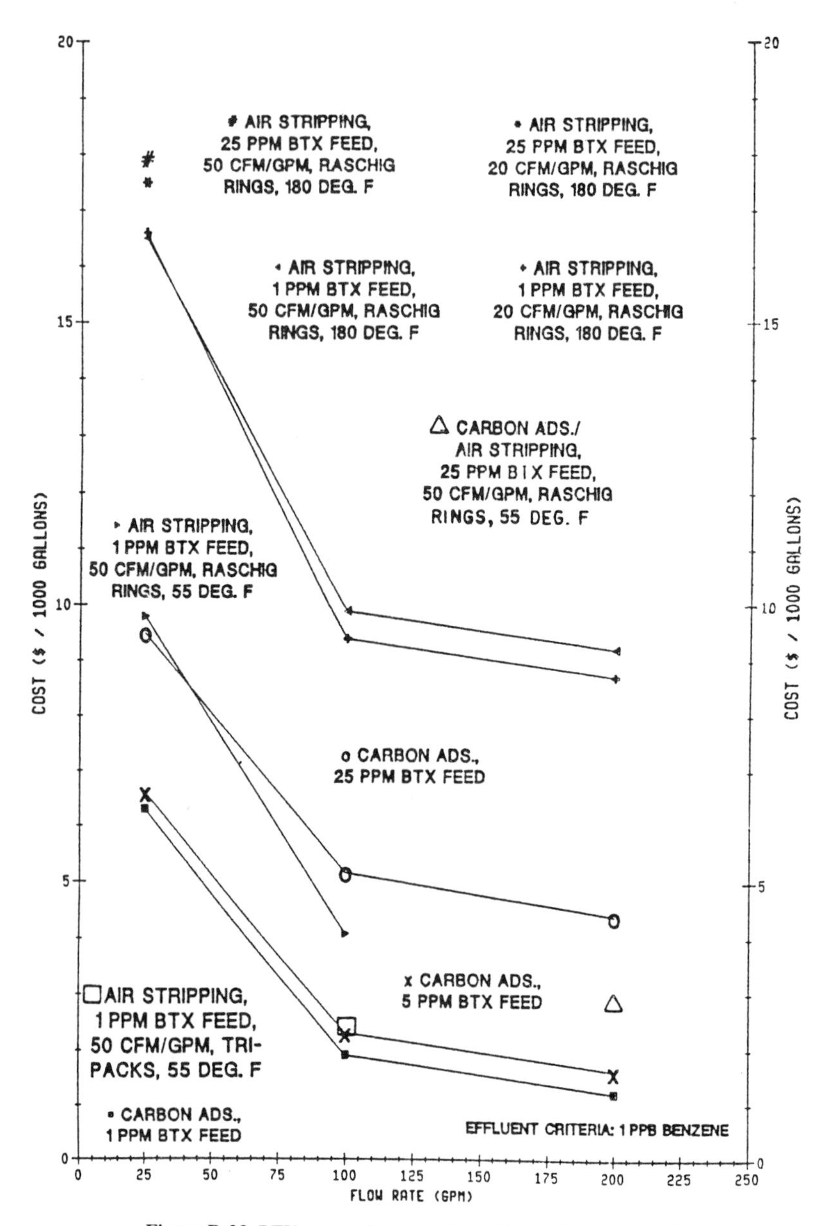

Figure B.23 BTX removal, comparison of technologies studied.

APPENDIX C

Soil Survey Map

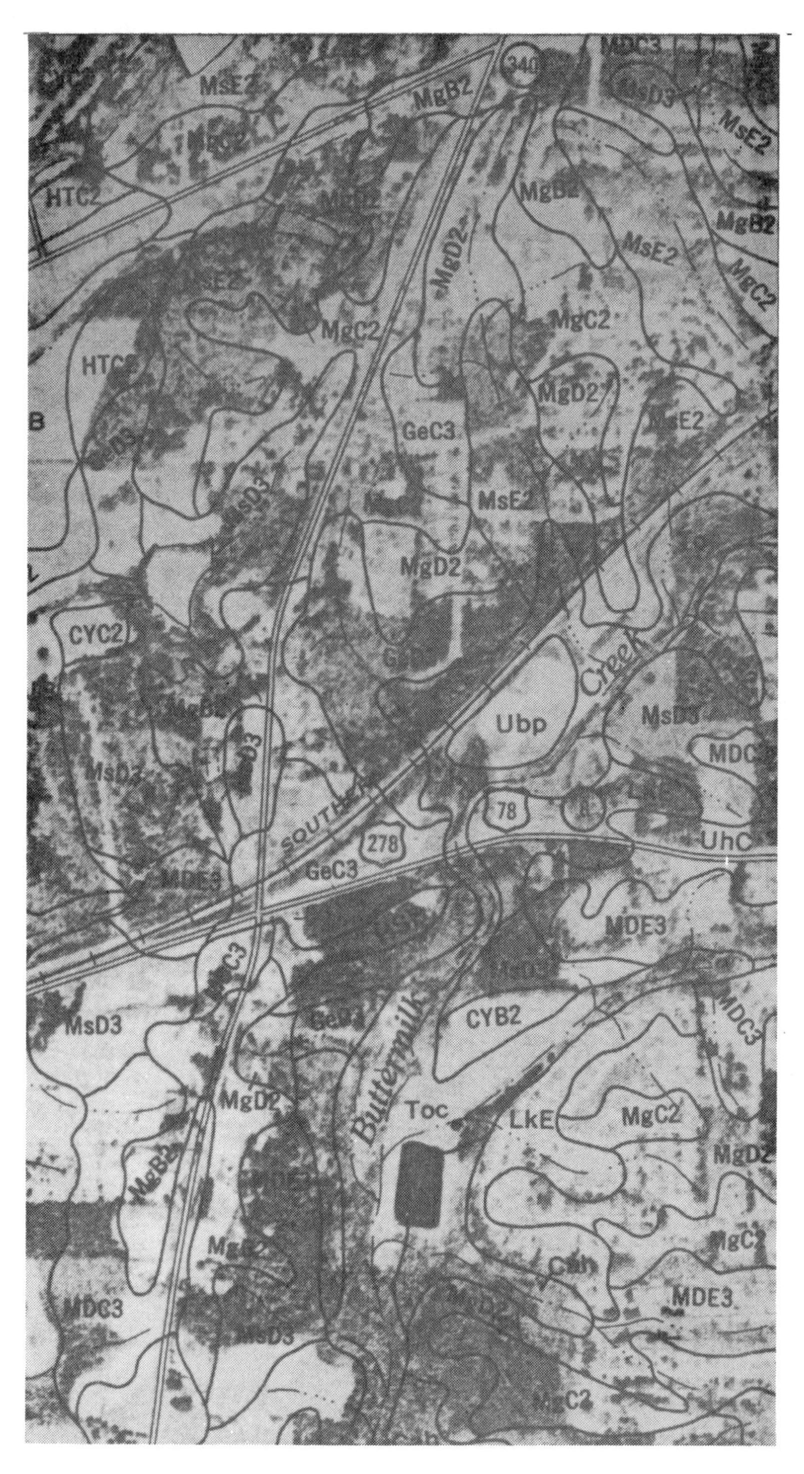

COBB COUNTY, GEORGIA

Sample of Soil Types from Map in Appendix C.

Soil Series and Map Symbols	Depth to—		Depth from Surface (inches)	Classification		
	Hard Rock (inches)	Seasonal High Water Table (inches)		USDA Texture	Unified	AASHO
Madison: MgB2, MgC2, MgD2, MDC3, MDE3, MsD3, MsE2	>60	>60	0-5	Sandy loam	SM, SC	A-2, A-4
			5-10	Sandy clay loam	SC,CL	A-4,A-6
			10-25	Clay	MH, CL	A-7
			25-36	Sandy clay loam	SM, ML	A-4, A-6
			36-48	Soft schist rock with less than 5% clay bands		

Percentage Passing Sieve—				Permeability (inches per hour)	Available Water Capacity (inches per inch of soil)	Reaction (pH value)	Shrink-Swell Potential
No. 4 (4.7 mm.)	No. 10 (2.0 mm.)	No. 4 (0.42 mm.)	No. 200 (0.074 mm.)				
85-100	85-100	60-85	20-40	2.0-6.3	.12	4.5-5.0	Low
95-100	95-100	80-95	40-60	0.63-2.0	.15	4.5-5.0	Low to moderate
95-100	95-100	85-95	60-80	0.63-2.0	.13	4.5-5.0	Moderate
90-100	90-100	65-85	36-60	0.63-2.0	.15	4.5-5.0	Low